Sustainable Development

This compelling collection of insights from seasoned civil engineering professionals sheds light on the crucial aspects of sustainable development through the unique perspective of those who help build our world.

The narrative begins by placing sustainable development in its rich historical context, examining how past events have shaped our current development trajectory and exploring potential paths forward within the framework of the 17 United Nations' Sustainable Development Goals. This book emphasises the devastating impact of emissions from unchecked development while highlighting the absence of effective worldwide mitigation strategies. It offers unconventional perspectives that may challenge conventional thinking—the views expressed are occasionally controversial, sometimes unconventional, and at times deliberately provocative.

While not designed as a traditional textbook or formal guide, this comprehensive volume serves as an invaluable resource and reference for development professionals and graduate civil engineers passionate about sustainable development. Its accessible yet authoritative approach makes complex concepts understandable without sacrificing depth or nuance.

Sustainable Development
Challenges and Assessment in Civil Engineering

Edited by

Scott Younger and Jasper Cook

CRC Press
Taylor & Francis Group
Boca Raton London New York

CRC Press is an imprint of the
Taylor & Francis Group, an **informa** business

Designed cover image: Clockwise from top right: (i) River ferry crossing in rural Guinea. Courtesy: J. R. Cook. (ii) Village rainwater collection, Bali. Courtesy: East Bali Poverty Project, https://www.eastbalipovertyproject.org/ (iii) Three generations of transport bridge across the Forth, Scotland. Courtesy: M. J. C. Younger. (iv) Glasgow University, Scotland. Courtesy: Glasgow University, https://www.gla.ac.uk/

First edition published 2026
by CRC Press
4 Park Square, Milton Park, Abingdon, Oxon, OX14 4RN

and by CRC Press
2385 NW Executive Center Drive, Suite 320, Boca Raton FL 33431

CRC Press is an imprint of the Taylor & Francis Group, an informa business

British Library Cataloguing-in-Publication Data
A catalogue record for this book is available from the British Library

ISBN: 9781032935270 (hbk)
ISBN: 9781032935249 (pbk)
ISBN: 9781003566274 (ebk)

DOI: 10.1201/9781003566274

Typeset in Times
by codeMantra

Contents

Foreword

The importance of the concept of sustainable development has become clearly established in the 21st century. The United Nations' sustainable development goals (SDGs), adopted in 2015, have provided us with a global framework for action.

The interconnected nature of the SDGs highlights the importance of considering sustainable development holistically. Also highlighted is the need for collective efforts among governments, businesses and individuals.

But as has been emphasised in this book, sustainable development is not a new concept. The fundamental need in the global economy to meet present population needs without impairing the ability of future generations to meet their own aspirations is a notion that has prompted serious reflection for many decades. Throughout the process of post-World War II economic development, the interconnectedness of economic growth, environmental protection, and social inclusion has been on the agenda of development experts. Ever since the 1970s, organisations such as the Club of Rome have pointed to issues around natural resource depletion in their publication *The Limits to Growth*.

Global challenges such as climate change, the destruction of biodiversity, and social inequality have, of course, worsened in recent years. This has further underlined the importance of sustainable development.

I am particularly pleased that much of what Dr. Scott Younger (an alumnus of the University of Glasgow) considers in this book in conjunction with his co-editor Dr. Jasper Cook and other contributors, resonates with the institutional values of his alma mater. Many of the challenges identified here are also in areas where the university currently plays a leading role in finding long-term solutions. This is evidenced by its top 20 global ranking in the QS World University Sustainability rankings and top 12 in the world in *The Times* Higher Education Impact rankings, which measure the contribution of a university in addressing the SDGs. However, this book adds the important perspective of the practitioner. Practitioners such as engineers, urban planners, environmental scientists, community organisers, and other development professionals play a vital role in setting out the challenges of sustainable development. Dr. Younger's and Dr. Cook's practical experience in many countries and reflections on their own experience and how they relate to historical and current perspectives on development provide a unique insight into how policy and practice have evolved. Practitioners interact closely with local stakeholders, governments, multilateral organisations, and NGOs. They provide a very useful perspective on a topic, which will remain central to our policy agendas for the foreseeable future.

Anton Muscatelli
Principal and Vice-Chancellor
University of Glasgow
Glasgow, Scotland, UK

Foreword

Throughout the long and arduous course of human history, progress was an extremely slow and sporadic process; perhaps only the classical period of Alexander the Great gave a time when knowledge and its transmission were properly valued. However, the real change in pace only took place over the past two and half centuries when the rapid rise in technological discoveries first took place. The 18th-century period is sometimes known as the Scottish Enlightenment, with such august names as David Hume and Adam Smith. It spread across Europe as the European Enlightenment. Towards the end of the 18th century, this enlightenment transformed into the Industrial Age through key technological developments.

This book is timely, a work that confronts the most pressing challenges of our era: sustainable development, climate change, and the parallel ethical responsibilities that must guide the future. In a time when the speed of technological progress would appear to outstrip our collective wisdom, this book's analysis serves as an important beacon.

The history of human civilization has been complex and often contradictory. For millennia, knowledge progressed slowly, but this book serves as a reminder that the future will face not just complicated, sometimes conflicting, technology but also profoundly existential paths. In these uncertain times, when the global political landscape is often marked by fragmentation, denial and inertia, this work stands as a critical thought and informed dialogue. It is a bridge across disciplines and between generations, inviting engagement with our shared predicament and finding pathways towards a just and sustainable future.

The old poet, Mak Dizdar, once urged, "When no one listens, write!". This book fulfils precisely such a call, not only speaking to contemporaries but also reaching out to future generations. I hope this book will inspire readers to join the collective effort required to heal the planet and restore harmony between humanity and nature. In this challenge lies not only our very survival but also the hope for a flourishing future for all life on Earth. The editors and contributors have provided an excellent and fulfilling coverage of this complex subject.

Anis Bajrektarevic
Swiss University, Vienna
Founder of the International Institute of Middle East and Balkan Studies (IFIMES)

Foreword

This book by Drs. Younger and Cook, with other named contributors, is a very timely contribution on the subject of sustainable development. This has become 'popular' since the United Nations (UN) adopted it in 2015 to describe the overall actions to assess the 17 goals that are required to judge whether a country or state is making progress. This book describes, in particular, the key issues that are important, such as governance, education and poverty, the state of the economy and crucially the environment. Climate change has been raised since it is an issue, which more and more occupies not only the UN concerns but also the concerns of citizens. Adaptation and mitigation for likely future changes are occupying minds.

After a general discussion of the key elements of sustainable development, the second part is slanted towards topics that often require the skill set of the civil engineer, such as energy and water, as well as transportation and shelter and industrial structures. As Drs. Younger and Cook rightly point out, much has changed over the past 60-odd years, perhaps the most marked being the acceleration of the growth rate of the population and the differing distribution of people. In parallel, the advent of the computer and the astonishing acceleration at which it can be called upon to resolve increasingly complex problems has led to today's Digital Age and on towards artificial intelligence.

In conclusion, Drs. Younger and Cook correctly state that it is not easy to predict sustainable development very far into the future, but perhaps, as the UN has done, measure progress in smaller bites, like 15 years. The authors are to be congratulated for this thought-provoking work.

S. D. Darmono
Chairman of the Jababeka Group, Jakarta, Indonesia
Founder of President University, Jakarta, Indonesia

About the Editors

Scott Younger, OBE, BSc (Hons), MS, PhD., DEng, CEng, FICE, is a chartered civil engineer with over 60 years of experience as a professional civil/geotechnical engineer, academic, consultant, and strategic advisor. He served as a lecturer in geotechnical engineering at Strathclyde University for seven years from 1965, followed by his association with World Bank, Asian Development Bank, and United Nations Development Programme on infrastructure and academic projects in Southeast and South Asia for more than 40 years. He supervised the construction of new provincial highways in Thailand between 1978 and 1983. He was an associate professor at the Asian Institute of Technology (1980–1983), where he also served as the editor of South East Asia *Geotechnical Engineering Society* journal. Between 1986 and 1992, he oversaw the establishment of the MSc programme in Highways and Transportation at the Institute of Technology, Bandung which won the 1992 IBM award for Sustainable Development. Founded Jakarta branch of Inst. Civil Eng in 1996. He also served as the Chairman of Strategic Intelligence (1998–2005) and Director of Thames Water Indonesia (1998–2002). He was a board member of the British and European Chambers of Commerce and Vice-Chairman of the International Chamber of Commerce. He has been actively involved in projects concerning rehabilitation and reconstruction of Aceh following the December 2004 tsunami disaster. He is the author of many papers on aspects of geotechnical engineering, water and highway engineering and poverty alleviation. He has served as a columnist for magazines like *Globe Asia* and *Forbes Indonesia.* Between 1973 and 1977, he was also active in politics and stood as a parliamentary candidate in the 1974 elections. He is currently an Honorary Fellow of Glasgow University.

Jasper Cook, BSc (Hons), MSc, PhD, CGeol, FGS, MCIHT, is a chartered geologist with over 50 years of experience as an engineering geologist/geotechnical engineer specialising in construction materials, geohazard and risk assessment, climate-resilient infrastructure development, research management and knowledge transfer programmes. He has spent over 35 years on infrastructure development programmes with UKAid-DFID, Transport Research Laboratory (UK), World Bank (WB) and Asian Development Bank in Africa and Asia. He was also a visiting senior research fellow at the Department of Civil Engineering, University of Strathclyde. Recently, he served as Research Manager and Team Leader on a multi-million-pound applied research programme for UKAid-DFID in 15 countries across Asia and Sub-Saharan Africa. He has been a consultant to the WB on climate resilience and road engineering issues for multiple transport projects in Sub-Saharan Africa, South Asia and Southeast Asia for nearly 15 years. He is the author of a wide range of peer-reviewed papers on engineering geology, geotechnics and road engineering. He co-authored the book *Rural Road Engineering in Developing Countries* (2023) along with Robert Christopher Petts.

Contributors

David Booth
East Bali Poverty Project
Bali, Indonesia

David Boyd
Stirling, Scotland, United Kingdom

Jasper Cook
London, United Kingdom

Hamish Goldie-Scot
Duns, Scotland, United Kingdom

Euan Low
Seoul, South Korea

Azra Meadows
School of Biodiversity, Health and Veterinary Medicine
University of Glasgow
Glasgow, Scotland, United Kingdom

David Meigh
Jakarta, Indonesia

Mark Murray
Research Centre
President University
Bekasi Regency, Indonesia

David Parry
PT Nusantara Infrastructure
Jakarta, Indonesia

David Wignall
England, United Kingdom, and Indonesia

Scott Younger
University of Glasgow
Glasgow, Scotland, United Kingdom

Acknowledgements

The editors would like to thank each of the following people for giving their time and expertise in the making of this somewhat complex book and undoubtedly improving it. Thank you all once again.

David Booth, MBE, BSc (Hons), MSc, is a civil engineer. He is the Founder and Chief Executive of the East Bali Poverty Project. It is rated as one of the best NGOs in all of Indonesia. This project has lifted 17,000 people out of abject poverty, reducing 80% illiteracy rate to a level where some of their top students have become university graduates. Other businesses have been started and are thriving. David previously worked in Nigeria, Kuwait and England prior to three decades in Indonesia.

David Boyd, BSc, CEng, MICE, is a chartered civil engineer with over 50 years of experience with consultants and in private practice, and is a geotechnical engineering specialist and consultant civil engineer with interest in nuclear. Thirty-six years were spent with continuous involvement in the nuclear industry submarine facilities, new build and de-commissioning. He currently serves on the Experts and Adjudication panel at ITER in France and as advisor to EDF on the new Sizewell C Nuclear Power Plant in the UK. He has provided invaluable nuclear insights.

Hamish Goldie-Scot, MA (Cantab), MSc, CEng, FICE, is an acknowledged expert in understanding and managing infrastructure-related corruption risks, a specialist advisor over corruption issues, founding member of UK Anti-corruption Forum, director of GIACC-UK and adviser to CoST and MCDF. He has worked in over 50 countries and been resident in 6, variously in private sector, government sector and civil society roles. He is also guest lecturer at the University of Cambridge.

Euan Low, BSc (Hons), PhD (Cantab), CEng, CScientist, has practised internationally in bringing sustainability into the core of financial decision-making in a wide range of infrastructure projects. He worked at Mott MacDonald, PwC and the Green Climate Fund advising governments, international development and commercial banks, developers, private equity and international financing institutions.

Azra Meadows, OBE, MSc., PhD, CBiol, FRSB, FLS, ZSL, is an honorary senior lecturer at Glasgow University in the School of Biodiversity, Health and Veterinary Medicine, with specialisation in environmental biology. She has worked on many development and poverty projects in Pakistan for over 30 years while on staff at Glasgow University with her late husband, Peter Meadows. She continues to work on projects in Pakistan.

David Meigh, BSc (Hons), MSc, CEng, MICE, member of Comm Irrigation & Drainage (UK), worked for many years for UK consultants, but now he is an independent consultant undertaking projects for the WB and other lending agencies in the field of water/irrigation. He has worked overseas for three decades. Based in Indonesia, he provided inputs for hydroelectric energy.

Mark Murray, MSc, spent much of his adult life in Indonesia mostly working in finance. He has been the Chief Financial Officer for multinational and local companies and the Audit Committee for a public company. He has worked in a variety of fields—Oil & Gas, Mining, Renewable Energy, IT, etc. Since 2020, he has been appointed at President University's Research Centre, consulting on VCS Carbon projects, and serves on the Board of an infrastructure group.

David Parry, BA (Hons), MSc, PhD, DipEd, FISoil Science, FIPSS, FRGS, after a spell at Oxford, where he gained a teaching qualification, taught briefly before working with Hunting Surveys in several countries in Africa and then for Mott Macdonald in Indonesia where he became the Resident Director. After retiral, he took up the position of Chief Environment Officer of infrastructure company PT Nusantara Infrastructure, a position he still holds. He is an environment and climate specialist giving invaluable inputs.

David Wignall, BSc, CNI, MIHT, MCIT, has served as a chief ports engineer, as research company Director and as Chairman of a major container terminal in a 40-year-long career in the ports and shipping sector. At present, he is making a major contribution to ship safety and the radical improvement of the environmental performance of the shipping sector, as it seeks to transition to low- and zero-carbon fuels.

Thanks are also due for comments from Dr G. Benton (town planner/arch expert), Chris Kirwan (AI expert), Dayana da Silva (IFIMES, Brussels) and Michael Norton (independent water consultant).

List of Abbreviations

AASHTO	American Association of State Highway and Transportation Officials
ACTS	Accountability Capacity Trust Setting
ADB	Asian Development Bank
AfDB	African Development Bank
AGR	Advanced Gas-cooled Reactor
AI	Artificial Intelligence
AIV	Aggregate Impact Value
API	Application Programming Interface
ARRB	Australian Road Research Board
ASEAN	Association of Southeast Asian Nations
BECCS	Bioenergy with Carbon Capture and Storage
BCE	Before the Common Era
BoQ	Bill of Quantity
BWR	Boiling Water Reactor
CAGR	Compound Annual Growth Rate
CCUS	Carbon Capture, Utilization, and Storage
CE	Common Era
CO_2	Carbon Dioxide
CoST	Infrastructure Transparency Initiative
COVID	Coronavirus Disease
CSP	Concentrated Solar Power
DFID	Department for International Development
dwt	Deadweight (tonnes)
EBPP	East Bali Poverty Project
EIA	Environmental Impact Assessment
EOD	Environmentally Optimised Design
EPR	European Pressurised Reactor
ERM	Enterprise Risk Management
ERP	Enterprise Resource Planning
ESF	Environmental and Social Framework
ESG	Environmental, Social, and Governance
EU	European Union
EV	Electric Vehicle
FAO	Food and Agriculture Organization of the United Nations
FED	Final Engineering Design
FHWA	Federal Highway Administration (USA)
FIDIC	International Federation of Consulting Engineers
FTSE	Financial Times Stock Exchange

GAM	Islamic Separatist Group
GDP	Gross Domestic Product
GHG	Greenhouse Gas
GICs	Geomagnetically Induced Currents
GIS	Geographical Information System
GPP	The Global Procurement Partnership
GPS	Global Positioning System
GW	Gigawatt
HFL	High Flood Level
HLW	High-Level Waste
IAEA	International Atomic Energy Authority
ICE	Institution of Civil Engineers
ICF	Inertial Confinement Fusion
ICS	International Chamber of Shipping
IEA	International Energy Agency
IFC	International Finance Corporation
IFE	Inertial Fusion Energy
IFI	International Finance Institute
IFRS	International Financial Reporting Standards
ILW	Intermediate-Level Waste
IMF	International Monetary Fund
IMO	International Maritime Organization
IPCC	Intergovernmental Panel on Climate Change
IPF	Investment Project Financing
IRR	Internal Rate of Return
IT	Information Technology
ITER	International Thermonuclear Experimental Reactor
JICA	Japan International Cooperation Agency
KfW	Kreditanstellung für Wiederaufbau (German Development Bank)
km^2	Square Kilometres
kW	kilowatt
kWp	Kilowatt-peak
LIC	Lower Income Country
LLW	Low-Level Waste
LMIC	Low- to Middle-Income Country
LVRR	Low-Volume Rural Road
LVT	Low-Volume Transport
MCF	Magnetic Confinement Fusion
MDB	Millennium Development Bank
MDG	Millennium Development Goal
MIT	Massachusetts Institute of Technology
MSCI	Morgan Stanley Capital International
MTF	Magnetised Target Fusion
MW	Megawatt
NASA	National American Space Administration

NDC	Nationally Determined Contribution
NGO	Non-governmental Organisation
NIA	Nuclear Industry Association (UK)
NPV	Net Present Value
NSEDP	National Socio-Economic Development Plan
NSSS	Nuclear Steam Supply System
OCDS	Open Contracting Data Standard
OECD	Organisation for Economic Co-operation and Development
OPEC	Organization of the Petroleum Exporting Countries
PHWR	Pressurised Heavy Water Reactor
PIANC	World Association for Waterborne Transport Infrastructure
PIU	Project Implementation Units
PPM	Parts Per Million
PSH	Pumped Storage Hydro
PWR	Pressurised Water Reactor
QA	Quality Assurance
QC	Quality Control
QCBS	Quality and Cost-Based Selection
R&D	Research and Development
RBS	Royal Bank of Scotland
SCRAM	(an emergency shutdown of a nuclear reactor)
SDG	Sustainable Development Goal
SGDIF	Scottish Government International Development Fund
SME	Small Medium Enterprise
SMR	Small Modular Reactor
SNF	Spent Nuclear Fuel
SRI	Rice Intensification System
SuM4ALL	Sustainable Mobility for All
t	Tonne
TEU	Twenty-Foot Equivalent (containers)
TNI	Tentara Nasional Indonesia (Indonesian Armed Forces)
ToR	Terms of Reference
TRL	Transport Research Laboratory
TWh	Terawatt-hour
UK	United Kingdom (of Great Britain and Northern Ireland)
UKAEA	UK Atomic Energy Authority
UKAid	United Kingdom Aid
UN	United Nations
UNCTAD	United Nations Conference on Trade and Development
UNICEF	United Nations International Children's Emergency Fund (United Nations Children's Fund)
UNODC	United Nations Office on Drugs and Crime
USAID	United States Agency for International Development
VLLW	Very Low-Level Waste
WB	World Bank

WHO	World Health Organisation
WLRC	Whole Life Road Cycle
WNA	World Nuclear Association
WWI	World War I
WWII	World War II
WWF	World Wildlife Fund

Introduction

The idea for this book on sustainable development came about in the period from 2006 onwards while Professor Younger was reporting on developments in infrastructure to Chambers of Commerce Boards, British then European and finally the International Chamber of Commerce.

The current concepts on sustainable development are generally seen as emerging from the publication of what is known as the 'Brundtland Report' (UN, 1987). In the introduction, Brundtland stated the first aim of the report as being "to propose long-term environmental strategies for achieving sustainable development by the year 2000 and beyond". This identification of sustainable development as a key international target paved the way for the adoption by the UN of, firstly the Millenium Development Goals (2000–2015) and the subsequent, more comprehensive 17 Sustainable Development Goals (SDGs 2015–2030).

This book seeks to provide a view of sustainable development as seen through the eyes of a number of seasoned professionals. The views expressed in this book may in places be controversial, not be always conventional and may occasionally be provocative. They are, however, the collated views of some hundreds of years of relevant worldwide practical experience in the field of civil engineering and its associated specialisations.

An initial scene setting introduces the concept of sustainable development and identifies key areas that influence our ability to achieve sustainable development targets. Sustainable development is put into a historical context, focusing primarily on the major events since the start of the Industrial Revolution in Europe. A comment is made on how these events have shaped our development path to date and how it might move forward.

The two principal themes of Chapter 2 concern governance and the other side of good governance, corruption. General forms of government, from the dictatorial, authoritarian to the highly democratic and participatory, are noted, and relevant sustainability issues are highlighted. It is acknowledged that even open democracies are not free of corruption, and strategies are outlined for reasonable and effective ways of curbing it. As an example, the Infrastructure Transparency Initiative (CoST) is presented as a proven system developed to curb corruption on infrastructure projects and aid good governance.

Chapter 3 is concerned with the worldwide status and interaction of education and poverty in the context of UN sustainability goals SDG 1 and SDG 4. The key point is that there is a direct relationship between poor education and poverty. Several aid-related projects are used as case studies to exemplify the key points and provide lessons for ways forward to reduce poverty and improve educational standards. These include the East Bali Poverty Project and examples from Pakistan and Aceh Province (Indonesia).

An appreciation of the Earth's composition and history is important for understanding the relationship between sustainable development and our environment. In Chapter 4, the restless nature of the Earth is emphasised, and natural hazards resulting from this restlessness are discussed in terms of their impact on the sustainability of future development. Key issues on climate change and its mitigation are summarised. Important issues relating to the sustainable management of the upper layers of soil are identified and discussed.

Chapter 5 reviews the key aspects of the urban and rural environments within which sustainable development has taken place in the past and which must do so in the future. This review encompasses the enormous demographic change in world populations within the last century in terms of both size and the rural–urban balance. The shift from rural to urban environments is now a worldwide phenomenon that raises significant challenges to sustainable development. Food security, pollution, climate impact, the need for 'smart cities', and new approaches to agriculture are all seen as issues to be addressed if SDGs are to be achieved.

Since the first harnessing of wood fire in the earliest Palaeolithic times, fossil fuels have been at the heart of human development. This dependence, particularly on coal, rapidly accelerated from the end of the 18th century. With the gradual expansion of the Industrial Revolution came increasing energy needs, with gas and oil supplementing coal from early in the 20th century. However, the effect on greenhouse gases (GHGs) was only noticed once the population of the world really increased significantly from the 1960s. Chapter 6 reviews fossil fuels, principally coal, in terms of historical and current increasingly unsustainable use. Alternative clean energy sources are listed, with the nuclear energy option selected for detailed discussion in terms of positives and challenges. Nuclear fusion rather than fission was highlighted as a likely principal energy solution for future sustainable development.

The water sector in relation to sustainability is dealt with in Chapter 7 from a global perspective. The importance of water is emphasised as being a vital resource for all life on the planet under key headings: agricultural needs, urban development, and rural development. Management of water is reviewed with examples from contrasting countries. Other key issues considered include land ownership, pollution, climate change impacts, and increased drawdown of groundwater levels. This latter issue is highlighted as a situation that requires immediate attention in light of the SDG requirements for poverty reduction and shared global prosperity.

The aim of Chapter 8 is to review aspects of sustainable transport as a provider of services and infrastructure for the mobility of people and goods. The subject of transport as a whole can include road, rail, sea, and air; however, this chapter focuses on sustainability relating to road and sea transport. In both cases, the historical trends in road and sea transport are reviewed as a basis for identifying current and future challenges, particularly as regards pollution and GHG emissions. Environmental aspects of port development are assessed alongside the needs of maritime freight transport. The questions of air and rail transport are mentioned briefly, particularly from the point of view of their impact on GHGs.

Chapter 9 presents a broad overview of disciplines within the field of civil engineering, not covered in Chapters 6–8, such as human habitations and major structures. Civil engineering as a whole is reviewed in terms of its influence on sustainable development of the future as well as the problems that are faced today. The significance of artificial intelligence (AI) is acknowledged as a natural follow-on from the earliest days of the thinking behind computer development. The combination of computing alongside AI is seen as a fundamental tool to be used in civil engineering as well as virtually all disciplines. It is becoming more important as each year goes by.

Chapter 10 explores the issues around understanding and assessing the financial sustainability of civil engineering projects and asserts that conventional financial performance reporting does not adequately capture sustainability risks and opportunities. Current financial decision-making is predominantly based around a retrospective review of financial performance and that, consequently, a nation's limited resources are not optimally allocated for society's long-term sustainable development needs. Important issues around financial sustainability are reviewed, and a selected number of sustainability financial reporting tools are reviewed in the context of working towards SDGs. In summary, the chapter states we cannot expect that simply using the financial sustainability standards will bring about all the necessary behavioural changes but, if at least, we can measure and report on how efficiently resources are being allocated and understand at least how far we are from a resilient and sustainable pathway.

The above chapters are followed by a summary in Chapter 11 that brings together their main themes and highlights some particular issues that are central to moving towards a sustainable future.

This book is not intended to be either a textbook or a formal guide or manual but a thought-provoking collection of views on key topics that seek to throw light on crucial aspects of sustainable development, which for the purposes of this book is defined as "The utilization and development of natural resources for economic growth with the exploitation of those resources to be held compatible with the sensible and sustainable conservation of the environment, such that future generations are not impoverished".

Looking back on the past 60 years, it can be seen in retrospect that the rather sudden growth in population from the 1960s, the impact of which was not fully understood at the time, was going to change the world in some significant ways. These changes are occurring in a manner that has not been realised fully yet. However, we are now fully aware of key changes, such as to the climate and the effect that GHGs, if not reduced or at least contained, would have on liveability and livelihoods. The trouble is that of the 200 nations of the world, some of the biggest GHG polluters are some of the largest. The other resource that is being altered is water, and this will have to be monitored closely. Whereas, in the not-so-distant past, the changes to life on the planet were relatively pedestrian compared with nowadays, and one could make forecasts for some decades ahead, it has now become quite a challenge to predict very far into the future, particularly since the population is still advancing at a significant albeit slowing rate, with a decline in numbers after about 2080. With these caveats, it is still considered worthwhile to look ahead with the best information available at a given time, noting that we have yet to fully process that which has already taken place in recent decades. AI will be increasingly applied.

Scott Younger and Jasper Cook
Editors

REFERENCES

Mahmutcehajic, R. (2011). *Across the River: On the Poetry of Mak Dizdar* (Abrahamic Dialogues). Fordham University Press.

Meadows, D.H,. Meadows, D.L, Randers, J. and Behrens, W. W. (1972). *The Limits to Growth.* Potomac Asso iates, Universe Books.

UN (1987) *Our Common Future (The Brundtland Report).* Report of the World Commission on Environment and Development. Available at: https://sustainabledevelopment.un.org/content/documents/5987our-common-future.pdf

REFERENCES

Setting the Scene

1

Scott Younger

1.1 INTRODUCTION TO KEY ISSUES

1.1.1 Definition

The term 'sustainable development' appeared in the early 1990s, and today, all development is around increasingly assessed in terms of 'sustainability'. Sustainable development is now commonly used as a yardstick to measure the validity of, for instance, engineering projects, to determine their life-term cost and thus their long-term value. Engineering projects, which are important, if not vital, to a sustainable future, concern those in energy, water and, to a great extent, transport.

A working and pragmatic definition of sustainable development for this book is proposed as follows:

> The utilization and development of natural resources for economic growth with the exploitation of those resources to be held compatible with the sensible and sustainable conservation of the environment, such that future generations are not impoverished.
>
> *(Younger, 2007)*

This definition attempted to cover the social, economic and environmental aspects of all life; it is assumed that the political aspirations behind human gatherings were essentially peaceful, not controversial and opportunities for aggressive behaviour. The concept was, if it were to be seriously followed, the creation of an environment where sustainable development strategies could and would be implemented. However, it has been found that it is impossible to ignore the political scene as this has played a major role over development or no development as the case may be. Sustainable development

DOI: 10.1201/9781003566274-1

had therefore to take into account four dimensions, namely social, economic, environmental and political issues (SEEP).

The term 'sustainable development' is applied in its broadest sense. The changes in the environment and climate are understood to be complex, many millennia in the making, and continually evolving. It is of value to look back at the historical past and review how we have arrived where we are in terms of development. It is useful to go over the different factors which take us towards a sustainable future and those which are a drawback to achieving this aim. There is a change, largely because of the recent significant population expansion and its impact on the use of resources, and in terms of issues such as waste and pollution. This will create unknown long-term consequences to the continuance of life as we know it and project it, and this requires more serious research than earlier imagined.

1.1.2 Sustainable Development Goals

The United Nations (UN) introduced the term 'sustainable' to apply to what they conceived as their development goals over the 15-year period, 2015–2030. They identified 17 key areas requiring attention, as listed in Table 1.1. These took over from the UN's Millennium Development Goals (2000–2015) and are monitored regularly through an annual reporting mechanism. They are expected to be updated again in 2030, with 'Sustainable' continuing to be the term used and thereafter revised every 15 years. Recently, there have been increasing efforts to attract businesses to take an interest in the UN Sustainable Development Goals (SDGs) through seminars and courses. This is considered vital for a successful outcome.

The 17 SDGs form an interdependent matrix with, arguably, a key group of seven issues guiding the way for the other SDGs, instead poverty, food, health, education, energy, water and sanitation (UN, 2025). Notably absent is any reference to the need for sound, educated and experienced governance, and addressing the crippling, endemic corruption of government leaders and big business officials.

These goals are particularly concerned with developing countries with limited or no access to food and water, inadequate or no sanitation, health issues and unsatisfactory education and survival opportunities. In particular, Africa remains a major target for UN SDG attention.

1.1.3 Population

The global population is forecast to be approximately 10.5 billion in 30 years, and based on previously accepted definitions of urbanised areas, the world's urban population is expected to increase from 55% in 2018 to almost 70% in the same time frame. Shelter, infrastructure and social services will be required for an additional 2+ billion families in less than 30 years across the world. With jobs. A tall order? It is doubtful that today's world leaders and politicians realise it, or can produce solutions to address the problem.

TABLE 1.1 Sustainable development goals (2015–2030)

NO.	*TITLE*	*TARGET DESCRIPTION*
1	No poverty	End poverty in all its forms everywhere.
2	Zero hunger	End hunger, achieve food security and improved nutrition and promote sustainab e agriculture.
3	Good health and well-being	Ensure healthy lives and promote well-being for all at all ages.
4	Quality education	Ensure inclusive and equitable quality education and promote lifelong learning opportunities for all.
5	Gender equality	Achieve gender equality and empower all women and girls.
6	Clean water and sanitation	Ensure availability and sustainable management of water and sanitation for all.
7	Affordable and clean energy	Ensure access to affordable, reliable, sustainable and modern energy for all.
8	Decent work and economic growth	Promote sustained, inclusive and sustainable economic growth, full and productive employment and decent work for all.
9	Industry, innovation and infrastructure	Build resilient infrastructure, promote inclusive and sustainable industrialisation and foster innovation.
10	Reduced inequalities	Reduce inequality within and among countries.
11	Sustainable cities and communities	Make cities and human settlements inclusive, safe, resilient and sustainable.
12	Responsible consumption and production	Ensure sustainable consumption and production patterns.
13	Climate action	Take urgent action to combat climate change and its impacts.
14	Life below water	Conserve and sustainably use the oceans, seas and marine resources for sustainable development.
15	Life on land	Protect, restore and promote sustainable use of terrestrial ecosystems, sustainably manage forests, combat desertification, halt and reverse land degradation and halt biodiversity loss.
16	Peace, justice and strong institutions	Promote peaceful and inclusive societies for sustainable development, provide access to justice for all and build effective, accountable and inclusive institutions at all levels.
17	Partnerships for the goals	Promote peaceful and inclusive societies for sustainable development, provide access to justice for all and build effective, accountable and inclusive institutions at all levels.

The unforeseen increase in the world's human population has resulted in exacerbating some problems of environmental despoliation and destruction, leading to unprecedented levels of migration and refugee exploitation. The demographics of society are changing. Humans are living longer. Female fertility is noticeably reducing in an increasing number of societies, particularly in wealthier countries, where the average age of populations has more than doubled from the premodern-day life expectancy of about 30 years to over 70 today. Even life expectancy in the DR Congo is 53 years, one of the world's poorer countries, while in Japan, one of the wealthiest countries, it is 83 years.

Just in the last 60+ years, population growth has reached unprecedented levels, as shown in Figure 1.1 (Younger et al., 2022), and there has been a realisation that sustainable development would require continual adjustment to meet the life challenges of the future. The population of the world had increased from 2.6 billion by the 1950s to over 8+ billion today, and it is forecast to increase at a decreasing rate to about 2075 and then reduce thereafter towards the end of the century, as indicated in Figure 1.2. Ritchie (2024) clearly explains the reasons for the population increase, given as follows: significant decease in child deaths due to health issues and mothers' deaths during childbirth in developing countries; decrease in child poverty; and significant increase in longevity because of better food, access to water and improved sanitation.

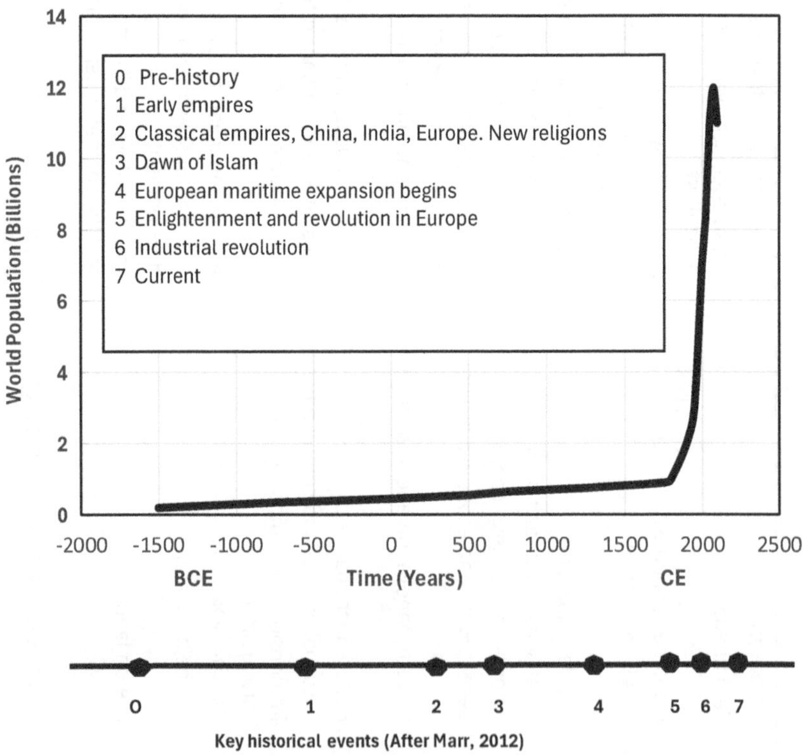

FIGURE 1.1 World population with time. (After Marr (2012).)

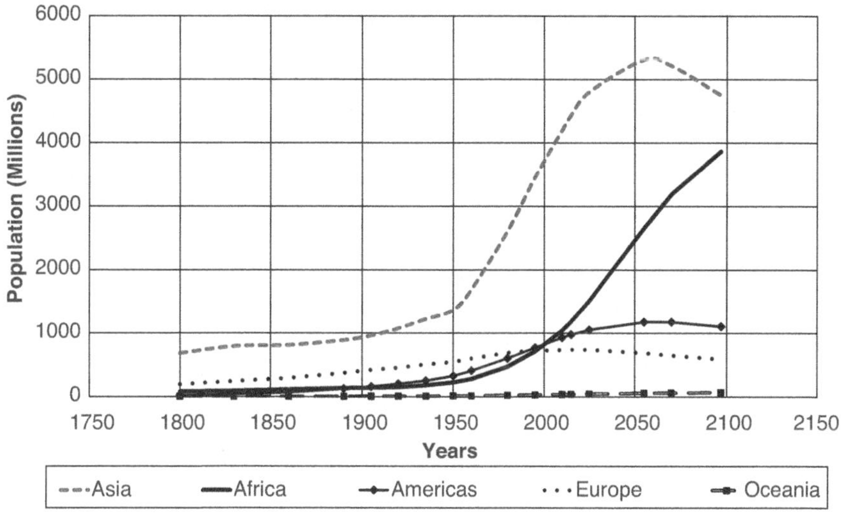

FIGURE 1.2 World population by continent 1800–2100 (Statista, 2020).

While today in developed countries, over 70% of the population live in towns and cities, 80% of new urbanisation is taking place in developing countries, notably Asia, and fairly imminently, Sub-Saharan Africa. Settlements have in the past taken many forms, similarly today, from the smallest hamlet to today's megacities. The dates of the earliest domestication are unknown. However, we do know remains have been found of an early tool-making area for anatomically modern humans at Jebel Irhoud Morocco (300,00 BCE). Later, a fishing-hunter-gatherer camp from the Upper Palaeolithic period, about 21,000 BCE were found at Ohalo II, now in Israel under the Sea of Galilee. Currently, more than half the world's population have moved from living in a predominantly rural environment to a predominantly urban one (UN, 2018).

By 2100, about 85% of 10.9 billion will live in cities. China and India will be accountable for many of this total, even allowing for their projected populations to decline from mid-century. China's numbers are already dropping. The present high birth rates in the developing world (Africa 4.5%) will decline due to loss of female fertility, largely due to increased city dwelling with its increased risk of disease contagion, better girls' education, a reduction in age difference between spouses resulting in a greater sharing of modern values, and net emigration.

Despite this, as noted previously, the continent of Africa's current population is projected to increase to a staggering 4.2 billion by the year 2100, about 38% of the world population – and about 10 times more than the declining population of Europe, projected to be 420 million in 2100. Nigeria will become the third most populous country in the world by 2050.

Meanwhile, the population of Asia will begin to decline by 2050, with China already having started this trend. Nevertheless, by 2100, the continents of Asia and Africa will hold 83% of the world population between them, very different from 1950 when it was 65%. Europe will have added little to the under 1 billion it supports today, and the Americas will signify not much more.

1.1.4 Government, National and Local

The governance of continents, nations and communities varies greatly in structure and aspiration. People's societies have created numerous forms of management by which they have chosen to be governed, or have been brainwashed, corrupted or forced into being governed. Governments can be essentially formed into democratic, non-democratic or other entities, which can be divided and further subdivided into different legislative authority systems.

The broadcast truth today is sceptically or even cynically received, and dismissed as at best half-truths, but too often just 'brainwashing'. Governments around the world are generally distrusted, with 'the people' getting more and more disenchanted with what has become common parlance in the UK, 'our political masters'.

But 'happiness' in today's world is so much to do with wealth and the state of the economy. This is framed around what form of government is in place – whether so-called 'democratic' or 'socialist', presidential, royalist, dictatorship or combinations of any of these.

A big question concerning the survivability of life systems, as climate change is taking place, is which form of governance is best suited for reaching SDG targets, for instance. Indeed, the question now is whether a pan-national form of government would take over the reins on all such matters on behalf of all sentient and non-sentient life!

1.1.5 Legal Issues

The serviceability and operability of legal systems have relevance to the effectiveness of SDG targets and related environmental matters. Experience has shown that more law is not necessarily better when it comes to protecting the fragility of the environment against a rich, corporate-backed media and its incentivising of lawmakers. The offering of inducements, whatever the kind, but suffice to say what corruption is to one, could well be a needed logjam-breaker to another.

Even with the best of intentions, extensive addressing of lengthy and complicated law does cause, all too often, costly delays to much-needed infrastructure, social facilities and low-income housing projects. The basic rural road, for example, of which there are millions of kilometres across the world, may require little or no political/legal inputs but they will have significant social and economic benefits (UN SDGs, 2015–2030; Cook & Petts, 2023). A major street project in an existing city, on the other hand, will likely have many political and legal hurdles to cross before the project can proceed. Any construction project, particularly if it is environmentally controversial, could take many months or even years to obtain the necessary approvals. This is certainly the case in the USA, UK and Europe. For any greenfield and brownfield development, once the basic planning laws and environmental regulations, as well as the opinions of the stakeholders, are dealt with, a start on work on the development, however, controversial, should be straightforward. With continuing changes in the economic, social and environment arenas, essential, even critical projects, bear witness to subsequent time delays, cost escalations and sometimes never see the end. There are no winners in this poor situation.

1.1.6 Investment in Development

The fact that there has been serious underinvestment in developing countries' infrastructure, with dedicated international aid funds squandered alongside systematic corruption from the leaders of these countries, is well documented. While corruption is not unknown in the developed world, sometimes to an unsettling extent, setting a clear example is proving extremely difficult if not impossible.

What will it take to allow sustainable investment in a nation's infrastructure? The concept of 'Nation twinning', such as that has happened between Scotland and Malawi, and very much in an embryonic stage with Jamaica, is seen as a 'win–win' situation. Here, financial aid is more effectively directed, and education, health, cultural and technical 'know-how' exchange programmes are carried out. This may be done in different ways by other twinning relationships.

A gauge of how each nation is performing in terms of the size of its economy and how its economy is performing is the 'Gross Domestic Product' (GDP). Its prime purpose is to help economists who advise governments (and extra-national bodies) whether a nation's economy is growing or experiencing a recession. Table 1.2 shows that the USA and China lead the table, except that an increase in a country's GDP indicates that, with the economy doing well, employment prospects will improve, and people should have more money in their pockets.

1.1.7 Social Interactions

Considerable financial aid is being asked by the Global 'South' from the richer nations of the 'North'. This is sometimes seen as 'conscience-ridden' reparations for the past few centuries of colonialism, occurrences that have led to environmental degradation, economic underdevelopment, racial profiling and poor social and structural infrastructure in many nations. This, in turn, has resulted in unequal access to healthcare education and social justice for the indigenous peoples. Strategies to encourage these fast-growing populations in the 'South' to remain and sustainably develop rather than migrating to the 'North' are of importance. It appears that large funding is urgently required in these areas, now suffering increasingly from both natural disasters and past and current social wrongs.

> War's environmental costs are steep … and in an era when armed conflict is on the increase, environmental harm has become 'collateral damage' that we don't even discuss.
>
> *(The Swaddle Team, 2023)*

Environmental damages caused by polluting water, destroying flora, poisoning soils, destruction of property caused by wars, displacement of people, and killing of mammals and birds lead to the prolongation of conflicts. An example is "the uprooting of 800,000 olive trees – a Palestinian agricultural mainstay – in the West Bank since 1967" by Israel (Swaddle Team, 2023). Other examples are the unquantifiable usage of war

TABLE 1.2 GDP of selected states and regions

RANK (2023)	*COUNTRY*	*GDP: MILLION US$ (2023)*	*GDP: MILLION US$ (2013)*
Top ten countries			
1	USA	27,360,935	16,843,190
2	China	17,794,782	9,570,471
3	Germany	4,456,081	3,733,804
4	Japan	4,212,945	5,212,328
5	India	3,549,919	1,856,721
6	UK	3,340,032	2,784,853
7	France	3,030,904	2,811,876
8	Italy	2,254,851	2,141,924
9	Brazil	2,173,666	2,472,819
10	Canada	2,140,086	1,846,597
Regions			
1	East Asia and Pacific	30,775,659	21,411,260
2	North America	29,509,296	18,696,254
3	Europe and Central Asia	27,240,629	23,454,504
4	Latin America and Caribbean	7,094,330	5,435,190
5	South Asia	4,475,568	2,389,729
6	Middle East and North Africa	4,287,684	1,624,047
7	Sub-Saharan Africa	2,032,749	1,820,439

matériel. According to UN estimates, "40% of intra-state conflicts in the last 60 years are linked to natural resources" (Swaddle Team, 2023).

The UN, as presently structured, stands accused of being ineffectual at best, when it comes to the eradication of interstate armed conflicts. This is unfair to the majority of UN members, since powerful nations, because of vested interests, sometimes exercise a veto and cause more harm when it comes to apparently sensible UN motions to intervene.

Women throughout history have not always been seen as inferior to, or subjugated by, men. Today, in some parts of the world, matriarchal or matrilineal societies do exist to various degrees. The Amerindian Six Nation Haudenosaunee Confederacy in the 1880s, where Amerindian women "had full and equal participation in their government, is based on the consensus with everyone having a voice", is compared with the time when white settler women had no rights and no vote. Referring to indigenous women's precedence over their men, Matilda Joclyn Gage, President of the Women Suffrage Association, was reported in the *New York Post* in 1875 as saying "Never was justice more perfect; never was civilisation higher"!

Tibetan Buddhist Mosuo people in Yunnan and Sichuan provinces in China have maintained a feministic governance over their society, with such as inheritances passed down to their daughters and divorces based on mutual respect. There is no marriage as such. In Sumatra, Indonesia, the Minangkabau culture is the largest known matrilineal society, with women's role as head of the household highly valued. Their selection of the chief and ability to fire him for failure could be a model to produce better governments elsewhere! In many societies, women are responsible for much of the farming, managing food supplies for the family. Women are generally excellent environmentalists.

Living longer brings with it not only the possibility of a more productive life but also the requirement for more care provision. As nations get wealthier in terms of dollars and cents, communities and families tend to disperse; the expectation is that time spent in education will get longer, and retirement conventions have to be reconsidered to maintain lifestyles.

1.1.8 Environmental Damage and Climate Change

There are recent signs and effects of climate change around the world; some uncomfortably close, for example, the recent increase in heatwaves, floods, droughts and wildfires, and some subliminal, as the increase in air and water temperatures, melting ice caps, rising sea levels, fossil-fuel-generated pollution and species extinction cause alarm to many. Most climate scientists agree that climate change is a cause for alarm to all life on this planet, and some see "global heating accelerating faster than is currently understood" (Milman, 2023). A relatively small number of voices dissent. In determining the significance to humans of climate change, increases in population can be important, probably the most important.

After several ever-increasing warming years, "the average recorded global temperature data in 2023 of 14.96°C was the highest on records going back to 1850" (Copernicus, 2023). The target limitation of 1.5°C beyond the 1850–1950 pre-industrial level was breached in early 2024, 6 years ahead of the ambitions of the 196 parties who signed the UN COP 21 Paris Agreement in 2015, a legally binding international treaty document on climate change.

Despite the climate treaty signatures, the built-in optimism of many, the educated exhortations of those who 'know', a growing number, especially among the younger generations, has grown sceptical of the political process. They understand the imposed imperatives of the status quo and believe that there will be little change. It is clear to the younger generation who will will inherit the environment, that more aggressive implementation of the changes will be required to maintain the climate within reasonable bounds and retain a liveable world for all sentient life now, which will be left in their hands.

The issues about sustainable development as influenced by climate change on urban conurbations are very different from those to be solved in rural areas. While taking up less than 1.5% of the world's total land area, "cities contribute over 70% of global CO_2 Greenhouse Gas (GHG) emissions, most of which comes from industrial and motorised transport systems that use immense amounts of fossil fuels and rely on far-flung infrastructure constructed with carbon-intensive materials" (Dasgupta et al., 2022). Urban planning, architectural and engineering designs through to construction material use must be thoroughly reviewed in terms of their suitability for the new 'clean air' city. Most countries of the world attending the COP 21 Paris meeting on Climate Change in 2016 agreed that the problem of global warming was due to greenhouse gas emissions (GGE). With the focus on South, South East and East Asian megacities – Delhi, Mumbai, Kolkata, Dhaka, Jakarta, Bangkok and Shanghai – where there is serious work to do, cities on every continent are trying with various levels of conviction and effectiveness to reduce their GGE.

One of the younger generation, who pulled herself out from a late teenage bout of doom and gloom, exacerbated by university lectures, is Dr Hannah Ritchie who, in her subsequent role as a data scientist, has written a book (Ritchie, 2024): *Not the End of the World,* in which she has analysed in some depth the key factors which affect the planet. As a tribute to the book, Bill Gates commented that it was "eye-opening and essential ... our planet isn't a lost cause..." The world will have to address the key issues; however, to achieve the necessary goals, and the political leadership of some of the main countries of the world must also show concern.

The climate has recently given a warning that the Earth is becoming hotter, and the UN and the majority of scientists attending the COP 26 meeting in Glasgow are ascribing this mostly to human actions. Obviously, the significantly greater population numbers mean increased activity and greenhouse gases (GHGs). Another signal is a change in the flow of ocean currents.

For example, the North Atlantic current which affects the UK and Europe; its circulation has an impact on the climate. There is a warning that the lack of action to contend with changes in the oceans will mean that there will be marked changes to water distribution and thereby agriculture in the future. This would have a deleterious, unsustainable effect.

Unfortunately, very few governments take the warnings for the future seriously enough and are postponing, for instance, all the necessary adjustments to their energy and water portfolios. Not enough is being done about reducing carbon footprints and waste and pollution levels. Virtually every country around the world is largely carrying on with 'business as usual'.

The target of net zero centigrade by 2050 will not now be achieved, not because it was impossible, but because some nations clearly believed that they had higher or more immediate priorities, sometimes politically driven.

In terms of the planet's 36.80 billion tonnes of CO_2 emissions in 2022, the highest emitter was the People's Republic of China with 11,397 billion tonnes (up from 8,621 bt in 2010), the USA with 5,057 bt (down from 5,680 bt) and the Republic of India's 2,830 bt (up from 1,677 bt), together making up about one half of the global total (Statista, 2022). The total CO_2 emission estimate for 2023 was 37.55 bt, a further year of increased carbon emission, largely from fossil fuels. Perhaps it is surprising given "the extreme events around the world would be sparking action but governments appear oblivious, somehow thinking treading water will deal with the flood of impacts" (Niranjan, 2023).

For other nations, the promised environmental solutions ought to be easier to action, given the will and delegated authority of the administrations to do what is needed to reduce individual contributions to these CO_2 emission figures. This will mean spending on physical assets, towards net zero, $275 trillion by 2050, or $9.2 t/year on average, an annual increase of $3.5 t. Also, it will mean not only job losses (est. 187 m), but also more job opportunities (est. 202 m). Implementations are already bringing net-zero-related investments' economic returns, something that is expected to improve in the future.

Serious efforts at effecting climate change become less manageable or achievable, when there is corruption involved, critically damaging the prospects of successful outcomes. Corruption is seen as a cost to society. It can be difficult to uncover, and the costs can be difficult to assess; it can be very significant for some government bodies, differing business dealings and on project budgets. Without condoning the practice, strict project management, a professional environment encouraging transparency is required when dealing with this issue. Recently developed in London, the CoST system is introduced in Chapter 2.

Serious issues which require a different mindset are the agricultural production industry and its pollution, and mining and the exploitation of limited natural resources. These are national government issues, but arguably, more importantly, give scientists of many disciplines, economists and environmentally active lay-citizens some concern.

An understanding of the sustainable challenges in the decades ahead is presented. Indices of navigating sustainability in infrastructure finance are discussed, looking at factors by which a society can view how it is faring and engages to improve. These are indices which must be continually updated as new research comes to hand, which is largely dependent on the availability of project development finance.

With today's increasing understanding of the regional complexity and unpredictability of the planet's climate, and the handling of inanimate, geological and extra-terrestrial environmental disasters, it will still be difficult, arguably impossible, to cover full structural or financial loss. It is important to be reminded that "there is a better than 95% chance that human-produced GHGs such as CO_2, methane and nitrous oxide have caused much of the observed increase in Earth's temperatures in the past 50 years" (IPCC, 2014). That happens to be the period that humankind's numbers greatly accelerated.

Brown (2021), has, from a political governance point of view, summarised quite expertly the immediate issues that face the world. These are climate change and the environment, especially the damage caused by human activities, and concern for global health. The threat posed by nuclear weapons is noted, as well as difficulties created by any international financial instability and opportunity inequality. He also raises the problem of global poverty and today's humanitarian crisis due, in part, to continuing population increase, and stresses the importance of education, shelter and opportunity. These issues will be covered when discussing the factors affecting sustainable development.

1.1.9 Education and Technical Development

Continued development has shown that larger investigations into today's environment and climate change causes and effects, could be attempted, modelled and implemented. Today, we are grappling with artificial intelligence (AI), software that follows human behaviour in tasks such as understanding speech and planning, enhancing, for instance, engineering design capabilities, and preparing deeper healthcare and education analyses. However, it generates its own rules, and that is where globally trustworthy solutions to problems must be created and regulations put in place.

Europeans, from those countries which have an Atlantic seaboard such as England, Spain and Portugal, began to explore the world about five centuries ago: the European Age, as Friedman (2009) has called it. They gradually put down markers, leading to later colonisation. Importantly, this led to the development of ships and shipping (Unger, 1994), which massively increased trade between different parts of the world. In turn, this encouraged the scholarly and curious minds of their day to research into new fields of science and included extensive work on developing energy sources, essential to fuel new industrial development.

The Industrial Age had its beginnings in the middle to latter part of the 18th century. It had a significant influence on the population and the way industry evolved and altered, in a relatively short time, the way that millions of people had lived up till then. At that time, the world population was almost 1 billion. In the past couple of centuries, new ideas and inventions have brought continual change to the meaning of 'sustainable development', a term as yet only intrinsically understood and practised by indigenous peoples.

There is the very important issue of education, as pointed out in the Institution of Civil Engineers (ICE) publication *Humanitarian Civil Engineering* (Kremmyda, 2021), which is mainly based on African experience. Education is a measure of how society is developed or developing, and how it can assist in sustainability, and a willingness to learn that humans cannot exist alone ignoring the rest of the biome in which they live. It is through education and understanding that people can keep on pushing the boundaries of learning and see how far civilisation can go without prejudicing the environment. Education, and the training and understanding that goes with it, is really the key to a society's future well-being.

1.1.10 Energy

Significant progress has been made by several countries towards the expansion of renewable energy and supporting customers with the shift, for instance, to electric vehicles, and heat pumps instead of gas boilers. "The world is on an 'unstoppable' shift towards renewable energy but the phase down of fossil fuels (use) is not happening quickly enough, (it is) double what it should be" (The International Energy Agency [IEA], 2023). China, for instance, despite extending its investment in solar, introduces new coal-fired power stations on a regular basis.

While science has it that fossil-fuel energy is a major contributor to Earth's present unacceptable level of GHG emissions, which must be reduced quickly, there will be for the foreseeable future the need for coal, oil and gas, albeit in steadily reducing quantities. It is unlikely that the needs of the road pavement, plastic and lubrication industries, for instance, can be met in total outside use of these fossil-fuel sources, but this requirement can be measured. For aviation, maritime shipping and long-haul trucking, and for such as the production of steel, cement and glass, requiring blast furnaces operating reliably at about 1,400°C, for the moment, economically viable alternatives to fossil-fuel use are not yet available or marketable. However, interesting strides are being made now across industries which will significantly replace fossil fuels in the future. Economics is not the sole justification when it comes to moving quickly in the direction of alternative energy sources.

The reality, however, is that a number of key suppliers, with vast, multi-trillion-dollar investments in extraction and downstream industries, armed with impressive financial and consequently political power with short-term vision, have generally little incentive to cash out and change. With oil (and natural gas) being plentiful, and the technology for extraction continuing to improve, making it more economic to produce and use (Gross, 2020), the intent to continue these lines of business, till supplies run out, prevails – or until 'green' solutions can be found as sensible and economic replacement. While more and more 'green' solutions are put in place, the pace of replacement is slower than deemed required, causing some concerns in the UN and some governments.

The critical issue of creating fossil-fuel-free energy, as already noted, in dealing with climate change, has been the fairly recent development and successful deployment of power generated from on- and offshore wind and ocean-wave and tidal turbines. Hydro power could be permitted in this mix of clean energies.

A clarification, if not a justification for the development and use of 'clean' nuclear energy and its longer-term future is discussed in detail, as the research on fusion, in place of fission, continues and will become commercially viable. Energy from hydrogen, preferring green to blue production, the incineration of biomass, and the rapid developments in solar, wind, wave and ocean tides are all real possibilities and will find some outlets.

1.1.11 Space

The exploration of space is over 60 years old when Russia and America entered the space race to put a man in space (Russian, Yuri Gagarin) followed by men on the Moon. In the intervening years, several countries have put many satellites up that circle the world for many purposes, some as bland as the weather but others for political reasons. Today, the National American Space Administration (NASA), is doing much research, including sending out further rockets into space, other planets, and exploring the other side of the Moon with private enterprise assistance. Other nations, India and China are also involved separately in Moon exploration, the latter having recently brought back samples of the surface for scrutiny, from the remote side of the Moon. It conjures thoughts that, at the century's end, the journey may be almost as common as taking a taxi! And NASA's exploration to Mars suggests that water can be found there, or even some planet moons.

One of the avowed purposes at the outset was that, through space exploration, the study of the Earth and the solar system would be seen from a new perspective, "advancing new technologies that will improve our daily lives, and inspire a new generation of artists, thinkers, tinkerers, engineers, and scientists" (NASA, 2023). This exciting extraterrestrial endeavour would "unite the World to inspire the next generation, make ground-making discoveries, and create new opportunities" (NASA, 2023).

1.2 'HARD' AND 'SOFT' DEVELOPMENT IMPACTS

In the context of development, a pragmatic division can be made between something that is produced, like a building or road ('hard' development), and something such as education or health improvement ('soft' development). They are frequently interdependent as may be seen by the links between SDGs three (Good health), six (Clean Water) and nine (Industry, Innovation and Infrastructure). Physical, 'hard' infrastructure elements and their maintenance are, in the main, the work of professional engineers, architects, urban designers, planners and builders. On the face of it, their responsibilities in carrying through building and civil engineering projects in a sustainable and environmentally positive manner are relatively logically and simply set out, albeit the project may be complex.

Furthermore, there are all the aspects of living that impinge on society, which present seemingly endless problems and challenges, such as climate, land movements and pollution. There are also, for example, the practice of legal and governance issues, political jockeying with seemingly endless disputes, which can be more complex and have a continual impact on sustainability.

1.3 DEVELOPMENT IN CONTEXT

1.3.1 The Industrial Revolution and Its Origins

In so many ways, our world has changed dramatically in the past two and a half centuries, but particularly and most importantly from the mid-20th century. Technical developments have moved along from the Industrial Age and entered the Information Age, "characterised by a change from traditional industries to an economy centred on information technology" (Wikipedia, 2025a). According to the United Nations Public Administration Network, the Information Age was formed by capitalising on computer microminiaturisation advances, which led to modernised information systems and internet communications as the driving force of social evolution. However, it is necessary to initially look at the period when the industrial revolution started, arguably with Adam Smith (1776), professor at Glasgow University, who published *The Wealth of Nations* in 1776, foundation of Economics, a decade or so before French labour revolted. From a more practical viewpoint, it is sometimes taken as 1783, the year in which James Watt developed the steam engine at Glasgow University. After 10 years of experimentation, he found a way to more effectively harness the power of steam, generated by the burning of coal or 'Black gold' as it has been called (Paxman, 2021).

The year 1815, the date of the Treaty of Versailles in Paris, is important in that it was when there was the realisation, after years of war, that there were significant socio-economic changes happening. From the engineering profession's viewpoint, it led to the establishment of a civil component for construction activities separate from military needs, and the setting up of the ICE in the UK in 1818.

The French Revolution is seen as a turning point in European and North American history. With its slogan 'Liberté, égalité, fraternité', the right to freedom, self-expression and equality was available to everyone. The growth in nationalism and the rise of the middle class are attributed to this singular event.

Friedman (2009) has called the five centuries till the end of the Second World War (WWII) the European Age. At the end of the 18th century, the second half of this period, the world population was just under 1 billion and took about a further century to add another 1 billion, reaching only about 2.45 billion at the end of WWII in 1945. This rose to around 3.0 billion in the early 1960s and then just and over 8.4 billion as of 2024. While there has been an increasing birth rate in some countries, in Asia, for example, people are living longer and some have healthier lives due largely to advances in medical science, better nutritious food availability, improved hygiene standards and more emphasis on education, especially for women. Contrast this rate of development with a few thousand years ago when the population of the world could be measured in 100,000s (Marr, 2012), and the benefits from important advances, like those which took place from the Stone Age to the Bronze Age over several centuries. Indigenous people on every habitable continent lived in various stages of development. These people today too often have been compared unfavourably with so-called 'advanced' civilisations, and the term 'primitive' persists, using comments such as 'Stone Age' to justify denying them their basic human rights.

A case in point is the comment made in 1961 by John F Kennedy who, in 'settling' the Dutch East Indies transfer of West Papua (Irian Jaya) to the newly independent Indonesian people, when he referred to the Papuans, a very different Melanesian people as "living, as it were, in the Stone Age" (Webster, 2009). With the politically opposed to the west USSR active in Asia, the USA felt that as there were only "7,000 Papuans and too primitive, they set the stage for Indonesia to absorb the territory, apparently with a clear conscience" (Rutherford, 2013). While the Dutch wanted to hold on to this last lingering remnant of the Netherlands Indies, Indonesia's first President, Soekarno, argued successfully to Kennedy and his advisors that nationalism, rather than a form of continuing colonialism, should prevail.

Consider the example of today's improving living conditions in countries such as Indonesia. From 1970 when the population was 100 million and the average longevity was only 40 years, its new government took the decision to concentrate on improvements to its rice-growing agriculture, provide better education, make school attendance compulsory up to age 15, and upgrade the health facilities nationwide. By 2000, the population had doubled and, most significantly, the average life span increased to 70.

An early key promotion of the development of the European Age, the movable-type printing press, invented by German inventor Johannes Gutenberg (died 1468), has had a profound effect on education and knowledge transfer. Then there were the inputs of the English polymath Sir Isaac Newton (1643–1727) who defined the laws of gravity and motion. He, separately, along with the German mathematician Gottfried Wilhelm Leibnitz (1646–1716), developed the usefulness of calculus in mathematics and sciences, both advancing the 10th century early work on calculus by the Persian mathematician and engineer, Muhammad Al-Karaji (953–1029). Further, there were the findings of other thinkers of the progressive German and French scientific schools in 16th and 17th centuries in the development of mathematics and physics, on which many subsequent inventions depended. For example, the English scientist, Michael Faraday FRS (1791–1867), is known for his important work on electricity in the 19th century.

Coal became the most important energy resource in the 19th century and the first two-thirds of the 20th century. Britain, through its natural resources of iron and coal, showed the way forward in industrial development. Germany followed from the later decades of the 19th century, then the USA, which benefitted hugely from the steady immigration of people crossing the Atlantic, escaping from general unrest, class struggles and poverty in European and other countries to the wide-open spaces that the USA offered.

1.3.2 The 19th Century Industrial Growth and Changes

Going back to the early days of the Industrial Revolution, James Watt's development of the steam engine in the 1780s from the previous simple pumping engines gave it a considerable impetus. Roads were also being upgraded and canals built from the mid-18th century but the railway and steam engine in the 1840s became the practical method of moving bulk goods. The journey times were obviously much shorter by rail than by canal.

The development of steam-driven shipping is closely linked to the outputs of three Scotsmen (Armstrong, 1975):

1. James Watt developed and first patented in 1769 significant improvements to rudimentary land-based steam engines that eventually allowed them to be considered for maritime propulsion.
2. Patrick Miller and William Symington developed the *Charlotte Dundas* in 1801: a steam engine-powered stern wheeler craft successfully trialled on the Forth and Clyde Canal.
3. Henry Bell designed the *PS* (paddle steamer) Comet. Launched in 1812, with a 4 hp steam engine capable of a speed of 5 miles/h. She operated successfully on the lower Clyde until she ran ashore and was wrecked in 1820. (The engine was saved for the Science Museum in London.)

The demonstrative success of the PS Comet (named after Haley's Comet of that time) led the way to the rapid development of steam power that led to Brunel's grand steam ship, the *Great Britain,* built in 1847, which travelled across the world, heralding an impressive increase in shipbuilding and the start of modern travel by sea.

A more significant advance at the time, however, was in rail transport. The best well-documented passenger journey by steam train was that of the Rocket, designed by an English civil and mechanical engineer George Stevenson (1781–1848), the renowned 'Father of Railways'. The train travelled between the towns of Stockton and Darlington in the north of England in 1829, just 23 miles (40 km), averaging 12 mph (19 kph) but capable of a top speed of 30 mph (48 kph). Less well-known, the inventor and mining engineer, Cornishman Richard Trevithick (1771–1833) had completed the first steam-powered locomotive to haul a load of coal on rails in 1804. The advances in railway engineering across the world, following on from this, were impressive throughout the 19th and 20th centuries. While the developments in railways were taking place, there were parallel advances in other aspects of civil engineering, for instance, for bridges, highways, dams and harbours.

The ability to move people and goods faster across continents and oceans boosted the growth of industry in Britain. This led to the expansion of existing settlements and the development of new towns and cities as entrepreneurs of industry now required office space, further places to live, shops, have their children educated, and an increasing number of additional social facilities. This would serve the rapidly expanding work force now required to develop, manufacture and market their inventions. The pace was rapid, and the British Parliament in Westminster was happy to let these early initiatives remain in the hands of entrepreneurs who would create this new urban environment and effect changes, good or bad though they were, with little interference.

In what was some of the most influential and far-reaching legislation ever passed, the government delegated extensive powers to elected local councils. These were dominated in the main by a middle class with a compelling faith in its own capacity to make the world a richer, cleaner, healthier, safer and better educated place. Some, however, through ignorance, venality or cupidity, or all three, were less altruistic.

1.3.3 Social Development

While the foundations of what we call the 'middle class' today are about five centuries old, it was only in the 17th century that it began to emerge as a force in society. The Glorious Revolution of 1688 in England, which ended the reign of the Stuart Catholic monarch James II and VII, saw William of Orange from the Netherlands, together with his wife, Mary Stuart (providing continuity with the in-place Stuart dynasty) accept jointly the English throne. This was on the condition that the system of absolute monarchy through the Divine Right of Kings was permanently abolished in England and Ireland, and the importance of rule by way of Parliament was established. Similar arrangements were enacted for Scotland.

In the UK by the end of the 19th century, 80% of the population of Britain lived in towns and cities. There were 141 towns and cities with between 20,000 and 100,000 inhabitants in 1861 (James, 2006). Twenty years later, the number had increased to 245. This was in line with a sharp increase between 1841 and 1901 in the UK's population from 26.7 million to 41.7 million over what became known as the Victorian Age.

These, as it is understood today, were the first stirrings of democracy in England. Without a written Constitution, however, there have been questions through to today as to whether Parliament in Westminster holds the right to legislate on some issues.

The emergence and rapid increase in the size of the middle class in the early days of the Industrial Age in Britain and elsewhere in industrialised Europe, such as witnessed in Paris, Vienna and Berlin, is largely attributable primarily to the development of new technologies, especially in manufacturing. Those that transformed the output of cloth manufacturing, for instance, especially in the north of England where a damp climate was important, were particularly noteworthy. At the same time, the opportunities to travel and move bulk goods were improved through the building and upgrading of canals, initially, and better roads, and then finally by the steady expansion of the railway network across and between countries all over the world.

The latter days of the Victorian Age saw the first stirrings of a social conscience throughout the Nation. Perhaps this was a natural follow-on from efforts by the English politician and philanthropist William Wilberforce (1759–1833) to abolish the iniquity of slavery worldwide some decades before. The large inflow of people to the cities needed a rapid building of homes and an infrastructure and social facilities to serve them. Urban building standards for the low-income citizens were too often utterly wretched; whole families were crammed into one or two rooms in shabby high-rise blocks with totally inadequate ventilation, and often many sharing sanitation facilities. The not surprising result was the creation of slum conditions and the spreading of disease responsible for long-term, poor physical and mental health of whole communities.

Over a century later, the worldwide human urbanising trend continues putting immense economic, infrastructural and administrative strains in the way of creating healthier habitation conditions. Any lessons learned from the UK's early urbanisation, 'social housing' in ill-serviced 'housing schemes', have been superseded by the sheer massive scale of the problems in many of the world's megacities. Known variously as shanty towns, bustees, skid rows, overwhelmed barrios, squatter settlements and even refugee camps, they persist. Despite deprivation in material terms, these inner-city ghettoes can contain thriving economies, around small-scale manufacturing and recycling. A few nations, such as China and Singapore, have made huge economic commitments to replace them.

A significant advance in the 19th century was the discovery that the terrible disease cholera was not an airborne miasma, but bacterial infection found mainly in polluted drinking water. It took some time for the medical profession to accept this finding. Glasgow was a fast-growing city and affected by disease, including cholera from insanitary wells and contaminated living conditions, essentially in the lower income high-rise homes (tenements). By 1859, the city had started to develop a supply of clean fresh water from Loch Katrine, which lies approximately 60 km to the north-west of the city. Designed by English water engineer John Frederick Bateman (1810–1889), this 88 m gallon (230 million litres)/day project was considered in its day to be one of the world's greatest engineering feats. The city still gets water from this source along the same route, with normal replacement and expansion necessarily taking place over the years.

Glasgow has changed shape significantly since the 1950s. Like many other British cities, it still wears the scars of decades of lack of due maintenance but has cleaned up its image from the neglected and polluted industrial city that it was. It is an excellent, recent example of a large urban 'clean-up', but there is still much to do in Britain and in most other countries too, improving both urban and rural environments. An alarming state of pollution exists, not just on land, rivers and other water bodies, and in the air, but also in the oceans. Environmental degradation has very significantly increased in the last 60 years. Ritchie (2024) has given some interesting solutions to a range of problems.

1.3.4 UK and the Far East

In East Asia, Japan was a country with a society which chose to keep very largely to itself. That is, until the mid-19th century, when involuntarily and abruptly under Western pressure and the effect of China's defeat by the British in a short, limited, war, it opened to trade with the rest of the World. One Westerner, who spent much of his life there, was a Scottish merchant Thomas Blake Glover (1838–1911). Glover befriended initially several of the factions who eventually became the leaders in the Meiji Restoration in 1867. He supplied them with guns and warships contrary to treaties and laws (He is remembered in the book, *The Scottish Samurai*). He also arranged in 1863 escapes for five and then in 1865 for 19 young scions of the samurai to go to Britain and attend institutions of higher learning. One, who attended Aberdeen University, founded Mitsubishi in 1883. Glover had warships built in Aberdeen, helped develop two coalmines, brought a dry dock, a small-scale steam train and cars to Japan, helping in some way for the Nation to set up a navy, a railway system, and other industries. The trade between Japan and Britain flourished, such that in 1894 the two countries signed an agreement, upgraded to a treaty in 1904, in part of which Japan agreed to assist in the looking after of Britain's interests in the Far East. This treaty was extended periodically such that Capt Brown, a friend of, Glover, was appointed Hon Consul for Glasgow and West of Scotland, which he handed on to his junior, A. Scott Younger in 1911, a post which was held for 20 years till retirement in 1931. By this time, the Japanese were developing a rather jaundiced view of the West, following poor treatment from Western allies in post-WWI treaty squabbles.

Japan was one of the Allies during WWI but was shabbily treated by the other members in the various treaty discussions. In 1923, the 1894 treaty was rescinded by Britain at the strong behest of the USA. The loss of face, of honour, will have hurt deeply, and could be one of the reasons why Japan took a strong stance against the West in WWII. The USA embargo on oil and gasoline exports to Japan in July and August 1941, to curb the latter's regional expansion aims, led to Pearl Harbour.

1.3.5 Influence of Marx and Engels; Early Days of Communism

No review of the direction of European politics at the beginning of the Industrial Age would be complete without reference being made to the German-born philosopher, economist and revolutionary, Karl Marx (1818–1883). Born in Trier, Prussia, he moved to London in 1849 as a refugee, to spend the rest of his life there. He had upset 'dyed-in-the-wool' leaders, first in Prussia, then moving onto France, where he studied the 1789 Revolution with interest, and Belgium, displeasing governments with his strong socialist views expressed vocally and in his journalism. He was supported in his arguably extreme views by another revolutionary socialist, the German philosopher, Friedrich Engels (1820–1895), who also had to help financially as Marx struggled with money throughout his life. Engels also wrote *The Communist Manifesto* (1848) jointly with Marx, who also wrote the three-volume *Das Kapital* (1867–1894). Marx foresaw the overthrowing of the bourgeoisie with the rise of the proletariat, arguing that the working class should carry out organised proletarian revolutionary action. To what extent he forecast the massive turbulence which would follow, the Great War (WWI), the execution of Tsar Nicholas II and his family in 1917 by the Bolshevist revolutionaries under Yakov Yurovsky, followed by the takeover of the Russian government by Marx's disciple, Vladimir Lenin, may never be known, but he undoubtedly was pointing the way.

1.3.6 Economic Development

During the past century, the USA outgrew all world economies, not just those of the European countries, by a very significant margin. By 2008, it had become by far the largest economy in the world, three times the next largest at that time, Japan, with having only about 4% of the total world population (Friedman, 2009).

Throughout the 21st century, the USA is expected to retain its prime position albeit less marked as the century advances (Friedman, 2009), assuming it overcomes its current internal social restlessness. Computer advances recently have been extraordinary, and with ongoing developments in AI (Kirwan & Zhiyong, 2024), for example, we wonder what it can do for humanity and the rest of sentient life. Would it help us solve these great environmental problems we live with today, and expect will get worse especially if our responses to the identified problems are slow? It will require much intelligent and judicial work to guarantee the technology that is developed will be used ethically and with probity.

1.3.7 Women in the Workplace: Gender Equality

Concurrent with these momentous and world-changing events, politically, with men essentially running the affairs of the nations of the world, things had to change, and changes to this patriarchal control were already being pushed into the background through various women's voting rights activities. From Sweden in the 18th century and other countries around the world in the 19th century, it took until 1893 when New Zealand became the first country to give women the right to vote at a national level. Early in the 20th century, the women's 'Suffrage Movement' agitated for these rights in the UK, but it was not until 1924 that they were given full voting rights.

Today, there is a growing realisation in most countries in the world that 'Women's Rights' are 'Human Rights', but across the world many women and girls face discrimination based on their sex and gender. "Gender inequality underpins many problems which disproportionately affect women and girls, such as domestic and sexual violence, lower pay, lack of access to education, and inadequate healthcare" (Amnesty International, 2025). "Gender equality and the empowerment of women and girls is not just a goal in itself, but a key to sustainable development, economic growth, and peace and security" (UN, 2025). According to Amnesty International, research has shown that all parts of society are improved when human rights are upheld and taken seriously. With figures showing that currently the median pay was 14.4% less for women than for men (House of Commons Library. 'The gender pay gap'. 8 January 2024), there is still much to do in this area alone.

The counter arguments against women's rights largely rely on beliefs that women and men are different, apart from the obvious. These claim that a woman's place is in the home, and men are the head of the household; women are too delicate to be exposed to (the corruption of) politics; that they are unable to understand the man's world of economic and political issues; and only a few radical, masculine and dominant women want to vote anyway. Feminists argue that religions act as a conservative force, maintaining the status quo, the patriarchal society. The French existentialist philosopher, writer and feminist activist Simone De Beauvoir (1908–1986) saw that religion was exploiting and oppressing women. She argued that religious faiths encouraged women to be meek, to put up with inequality, exploitation and suffering and doing so would bring rewards in the hereafter. Many religions give women the role of nurturing, care and giving birth. While this is generally applicable to the credo of western or Abrahamic monotheistic practices (where the implication is that God, after all is a man!), eastern religions and those of many indigenous peoples have a balanced view on this matter. In the Islamic faith, also an Abrahamic religion, the Quran states explicitly that men and women are equal before God. In the 7th century women could own businesses and fight beside their men in wars. However, the extreme theocratic countries, especially the Taliban in Afghanistan, don't allow women to do any meaningful work at all, losing half of the economic power base of the country. It is considered unsustainable even in the short run.

In the UK medical workforce, women today make up most of the doctors in general practice. The medical profession, as in the law, architecture and engineering, for instance, in virtually every part of the world, has in the past been dominated by men. Today, compared with just a century ago, there have been remarkable improvements in the status of women, but there is still a great deal to do to reach genuine and honest parity in the minds and actions of men, in education, and governments they elect. The UK Law Society reported that, since 1990, women represent over 60% of new entrants into the profession and there are now more women than men practising as solicitors. In architecture, women licensed architects are few, with reasons given for this disparity as the wage gap, lack of mentors, need for time for childcare, and the poor perception of present-day female leaders.

According to recent data there are 582 million entrepreneurs worldwide (Markinblog, 2025), of which 274 million (47%) were female (GEM, 2022). In the USA, 42% of all businesses are owned by women, and in 2021, 49% of all business start-ups were made by women, significantly up from 28% 2 years earlier. In the same year, it was noted that half of the entrepreneurs were women. In Latin America and the Caribbean, whereas in Canada and South Asia, they make up 35.6% and less than 20%, respectively. But figures vary hugely, determined by religion, and the level of women's education. While entrepreneurship in the West is opportunity-driven, in Asia it is much more necessity-driven. Women account for 41% of the global workforce and control >US$20 trillion in annual spending. There has been much more female participation in business over the last three decades but there is still much to change in this area.

1.4 RECENT DEVELOPMENT CONTEXT

1.4.1 End of World War I: Rise of Fascism and Communism

The first four decades of the 20th century, in Europe initially, were turbulent times, which perhaps inevitably, led to WWII. The dreadful carnage and serious maiming from battle wounds or disease of World War I (1914–1918) – 'he War to end all Wars' – had a dramatic effect on society on the continent and beyond. A peace agreement was signed at the Armistice in 1919, but disputations continued among the winning nations and a defeated Germany, which reacted badly to what it thought was unduly harsh treatment. Soviet Russia, later the Soviet Union, because of the Allies' fear of Bolshevism, was also kept out of the negotiations, as was their badly treated fellow ally, Japan. The result was uncertainty and social unrest throughout Europe. An exceptionally deadly global influenza pandemic (1918–1920) added a further 17–100 million deaths to those killed in the War. This was followed by the Great Depression (1929–1939), a severe, worldwide economic downturn which caused massive hardship in both rich and poor countries, with millions suffering high unemployment and widespread business failure.

The ideas and ideals of the Communists, which were being test-bedded in Lenin's Russia, were being challenged by a Fascist movement in Germany, flying the flag of an extreme racist, nationalistic and centralised autocratic political philosophy. However, it did not stop the German Reich and the Soviet Union, banished from the final treaty by the Allies at Locarno in 1925, signing the following year essentially a trade pact – 'The Treaty of Berlin' (German–Soviet Neutrality and Nonaggression Pact) which helped each of them recover. This was followed by a further treaty – the Molotov-Ribbentrop Pact – in 1939, guaranteeing peace between each other, with a secret protocol setting up basically an East–West boundary through Europe, defining each a sphere of interest. The German Reich, always with the intention of gaining *lebensraum* ('living space') in the East, back-stabbed Stalin's Soviet Union in June 1941.

The failure of the 1925 Locarno treaty, the cool reception by influential and powerful nations to the idea of a League of Nations (the forerunner of today's UN), the Great Depression, the rise of the proletariat, although not realised at the time, improvements in media communications, and general political and economic unrest, all led to the next serious global, brutal conflagration, WWII. It was estimated that almost 100 million men, women and children, across all continents, died as a direct result of these two wars. The consequential environmental damage was incalculable.

1.4.2 Invention of Television and Development of Media

Through all this, scientists and engineers were continuing their research and experimentation. One notable invention by a Scottish inventor, electrical engineer and innovator, John Logie Baird (1888–1946), was television, which first appeared in 1936. This led, in time, to entertainment and education becoming readily available to all income groups and a worldwide audience. The medium's capacity to educate as well as entertain has been nothing less than impressive and seen as being beneficial to the encyclopaedic advancement of humankind.

Through this medium, we are able to better understand such matters as its dominant, sometimes management role in the planet's environment – aquatic, grassland, mountains, forest, desert and tundra. However, it can be uncontrolled and is occasionally used mendaciously in the dissemination of information. Abuse of the medium through such as 'brainwashing' is causing concern. What will it take to have this medium and all its offshoots, such as computer screens, smart phones and watches to become less of a political tool and allow it to develop areas of science and differing cultures?

Governments worldwide are having difficulty in understanding, let alone managing even the simplest of social media. Sometimes, an unsuspecting public is subject to being fed unhealthy products which they cannot afford. "The whole scenario is in the hands of a tiny super elite … and that we have influence of any kind is sheer fantasy". (Euan Low, Personal Communication) Nowadays, the whole matter of an unmanaged AI requires much more attention.

1.4.3 Early Days of Travel by Air and into Space

The progress and expansion of ocean shipping has, over the past two centuries, been extraordinary, but it rarely figures in matters pertaining to the environment and the effect that it has on it. Around 90% of all global trade in 2023 was carried by ocean shipping. This was valued at US$2.2 trillion in 2021 and is projected to reach US$4.2 trillion by 2031.

The American aviation pioneers, the Wright brothers, Orville (1871–1948) and Wilbur (1867–1912), are generally credited with inventing, fabricating and flying the world's first successful powered aircraft between 1903 and 1905, although a decade earlier Clement Ader flew a very short distance in uncontrolled flight (Ader, 1906). This major step would eventually lead to worldwide air travel and the consequent rapid transfer of commercial and cultural developments. Over the following 50 years, the advances in aircraft technology were extraordinary (see Section 8.10.3). They went from early biplanes with wooden wings and canvas bodies, used in WWI, to propeller-driven and then onto the turboprop engines in the 1950s leading to commercial jet engines in the late 1950s and 1960s and even rockets to the Moon, and further to our solar system planets. Now with Voyagers 1 and 2, launched in 1977, travelling through the heliosphere into stellar space at over 63,000 km/h. As of 17 January 2024, the Voyagers had gone 24.4 bn km away from the Earth.

On space tourism, now the toy of billionaires but with a promised future for others, the question of pollution from fuel, human waste and debris has to be addressed.

The comment that the SpaceX Starship was a first experiment, largely 'a large tube filled with dollar notes, set fire to, and sent up into space,' where it exploded in a 'rapid unscheduled disassembly', but hailed as 'a success' (Guardian, 2023) is difficult to comprehend.

With the formalisation of the concepts of the algorithm and computation in the (1912–1954) period, these led to what can be considered as the general-purpose computer. It is widely recognised today as the forerunner of the theoretical computer and AI and its development has reached unimaginable complexity, with further developments to come. To date, it has seen the development of the internet and its offshoots, and from that flows instant information. It is hard to imagine now life without it.

1.4.4 Colonialism and Decolonisation

Colonialism has a long history, with its pluses and minuses in terms of the permanence of associated environmental and socio-economic damage, for instance. The British Isles from 55 BCE were settled by a succession of European peoples, including Romans, Saxons, the Vikings and Normans, and later to a smaller economic sense, French Huguenots and Dutch. Colonialism, in the modern sense, began with the beginning of the 'Age of Discovery', with the conquest of Ceuta in North Africa in 1415 by the Portuguese, the early days of the European Age (Friedman, 2009). The reasons given were "to control navigation through the Straits of Gibraltar, spread Christianity, amass wealth and plunder and supress predation on Portuguese populations by African slave-dealing Barbary pirates; at that point, a minor trade" (Wikipedia, 2025b). In the race to challenge the 'Silk Road', control of the trade routes to Asia, with improved nautical knowledge, early exploration of first Africa, then the Americas, was followed with further explorations in West (Middle East), South, Southeast and East Asia. By the 16th and 17th centuries, Portugal, Spain, England, France and the Dutch were establishing what became their overseas empires, in direct, and often aggressive competition with each other.

The first decolonisation took place between late 18th and mid-19th centuries when most of the European colonies in the Americas gained their independence. The British, French, Portuguese and the Dutch were turning their attention to southern lands of Africa and Asia at a rapidly increasing pace in what is known as the era of 'New Imperialism', a time coincidentally with Europe's increasing, resource-hungry industrialisation. Belgium, Germany and Italy were to join in Africa. As the pace of colonisation quickened, wars between the colonisers, in particular during WWI, led to colonies being (re)distributed among the winners. Unrest within these, sometimes with arbitrarily defined boundaries, ignoring tribal distributions, stirred nationalist feelings among the colonised. But decolonisation properly did not get underway until the end of WWII in 1945.

In the modern era, examples of quite different imperial nations which have disintegrated have been the Ottoman Empire, the Austro-Hungarian Empire, the Union of Soviet Socialist Republics (USSR), the Japanese Empire and the Kingdom of Serbs, Croats and Slovenes (later Yugoslavia). Meanwhile, others are currently expanding or threatening to expand their territories, such as The State of Israel into the State of Palestine, the Bolivarian Republic of Venezuela into the Co-operative Republic of Guyana, the

Kingdom of Morocco into the Sahrawi Arab Democratic Republic (Western Sahara) and the Russian Federation into Ukraine. Indigenous peoples and those lost to history in every continent continue their struggles for recognition. New nation numbers are now growing faster than those being lost. Today hegemony can follow traditional patterns of 'on the ground' territorial occupation but now can also be handled remotely. All can be carried out from a computer desk on the other side of the world. It is happening!

Prominent citizens in the larger colonised states were the Indian anti-colonial nationalist and social democrat Jawaharlal Nehru (1889–1964), 'Mahatma' Gandhi (1869–1948) and the Indonesian nationalist, orator and statesman Koesno Sosrodihardjo, better known as Soekarno (1901–1970) in the Dutch East Indies, later the Republic of Indonesia. Nehru and Soekarno had been educated in Europe in the 1930s and learned of the 'radical' thinking taking place, socialism and its more leftist form, communism. They joined robust street protests and resistance groups when they returned to their own countries.

As a direct result, 36 territories from Asia and Africa were given or took their freedom over a period of 15 years from the end of WWI (over a third from Britain), and they, in turn, joined the UN in their own right. From its 51 founding nations in 1945, there are now 196 member nations of the UN (2022), from the largest, the Republic of India with a population of 1.43 billion, to the smallest, the island of Tuvalu (ex. Gilbert and Ellice Islands) with 11,398.

In retrospect view of this past, 'New Imperialism Age's colonial accomplishments', a state of "self-consciousness (brought on) by a feeling of being critically commented on by others" (Rothermund, 2011), has arisen only now. Some nations tend to suffer from a general amnesia concerning their colonial past. Some, like Prime Minister Gordon Brown, said on a visit to Tanzania that "the times when the British had to apologise for their colonial rule are over". The presence of immigrants from ex-colonial countries in the respective metropolitan countries, has contributed to the significance of such debates.

1.5 TODAY AND TOMORROW'S WORLD

Once WWII was over and world peace restored, more localised but environmentally destructive wars took place – in Korea (1950–1953), ending with a division of the Nation, Vietnam (1945–1975) resulting in division and then eventual unification of north and south, Yugoslavia's breaking-up into pre-WWII countries, and the continuing, never-ending mayhem in the Middle East. Life for many people, however, appeared to have changed for the better in the 1960s.

Some young people, from about 1960, felt unfettered by the past and looked with increasing confidence to the future. They expressed this in more casual, colourful fashions and lively music. Over many parts of the world, there was a gradual levelling up of society and a significant improvement in living standards, certainly in the USA, UK and Western Europe, and Japan, but also eventually in Southern and Southeast Asia, China, and the developing world, latterly parts of India and some countries in southern Africa and South America i.e. much of the world.

For many others, life has not changed for the better, and famine, droughts, floods, disease, homelessness and a setting for further wars continue. Huge strides have been and had to be made in the medical field with the human population living longer. Increasing numbers of people can expect to live to and beyond 100 years. This will mean changes in the human experience and expectations, covering and upgrading the UN's SDGs. Namely

- New approaches not only to educating children from an early age, but also to the institution of lifelong education
- Improving lifestyle, nutrition and health
- Advancing and institutionalising gender equality
- Developing easily accessible sport and leisure opportunities for all – with shorter working and longer free time available
- Reassessments of retiral age psychology, developing concepts of community help
- Adjusting to the prospect of more job and vocation changes in a longer working life
- Management of adequate personal financial and care arrangements throughout life

People are already in several countries living where wellness and happiness are becoming the norm. All life, however, will have to adjust to changes, in time, to the climate and environment everywhere. Even with the likelihood of further wars of destruction, new diseases and pandemics, mass migration of humans and much of other life, the world will be quite different, before the 21st century has ended.

The later derivatives of the computer have been so extraordinary, life-changing, covering every aspect of life, that it has given rise to what was first called the 'Digital Age' in the 1970s with the introduction of the personal computer, and beyond to the 1 January 1983, the official birth date of the 'Internet Age'. The computer has gone through many iterations, from room-sized mainframes, supercomputers to desktop, laptop and wearable computers down to the smart phones and watches, and its coming use in AI. The young of this 21st century have a different set of important new and previously unknown opportunities and challenges on the horizon to address. Apparent entire, encyclopaedic information sources, and complex calculation capabilities are already changing the professional sanctities of the professions, particularly the medical, engineering, architectural, financial and legal.

Today, young children, barely 21-months-old in a child carrier, are seen fully absorbed with a parent's handphone or tablet, showing an expertise that their grandmother could never have imagined. Almost everything today's young do in their lives will have a link to a computer, however trivial the subject. Computing tools will be indispensable in virtually all they do in their lives.

In engineering and construction, the old, hand-drawn working drawings have virtually disappeared as computer graphics now produce construction documentation to a quality not dreamed of just a few decades ago. The potential of high-end, fully immersive virtual reality (VR) is largely used today for computer gaming, but is used very effectively, for instance, for architectural, engineering, urban design and landscaping presentation work. VR has been an educational tool in every educational establishment from schools to universities for some decades, and developments are moving apace. A top job in today's employment market is a computer engineer/technician/designer or data scientist; opportunities are expanding for these skills.

Despite these remarkable developments, with the recent increase in the world population, the number of those who are still living below the $2.04/day (in Ethiopia) and $24.55 (in the USA) poverty line totals approximately 8.5% of the world population (Our World in Data, 2022). The poorest in the world are not only undernourished, without access to such basics as electricity and safe drinking water, but also have little or no access to education, health provision and shelter. Those classified as in 'extreme poverty' numbered 660 million in 2019, increasing to 753 million in the following year, but declined to 680 million in 2022. Human destitution is very much a 'poor' country problem today.

One in three of the poorest 110 countries surveyed in 2023 by the Multidimensional Poverty Index halved their poverty rating in the last 15 years. According to this release, 1.1 billion out of 6.1 billion people surveyed live in acute multidimensional poverty across 110 countries; Sub-Saharan Africa and South Asia accounted for 923 million of them, and India accounted for 415 million of South Asia's total. However, China (69 million, 2010–2014) and Indonesia (8 million, 2012–2017) also showed improvement, and such successes are forecast to be added to in the coming years.

It is forecast that the global population in 2100 will be 10.9 billion or less (UN, 2019). While the decline in human reproduction will be seen mostly in the 'rich' and recently developed countries, the population of Africa is set to grow from today's 1.4 to 4.2 billion, particularly D.R. Congo, Nigeria, Tanzania, Sudan and Egypt. Three other countries are forecast to increase – India, Pakistan and the Philippines. China will have a steadily declining population from 2024 and India is due to follow two decades or so later. It will be a matter of time, perhaps next century, before all nations will show a decrease in population, but meantime humanity must find ways to survive, with the added responsibility of improving not only the lives of fellow humans but also all life on Earth.

Growing urbanisation is, to an appreciable extent, helping to alleviate poverty, raising education opportunities especially among women, but is adding crime to societies, overcrowding, 'slave labour' working conditions and the recreating of community units.

There are further social problems to address. The recent population growth and increased longevity have caused an imbalance in employment and job availability. The lure of urban life and the employment upskilling opportunities there attract the young. This has meant a dearth, particularly the young, of those remaining in the rural areas, changing the face of agriculture today. There is a return to similar conditions in the 19th century when people flocked to towns and cities, and rural areas were emptied of people. The problem is exacerbated this time, having to deal with far larger numbers, and the expanding use of robots, albeit in its early days, is already changing the job situation, making it more difficult.

Falling population growth is still to be established at sustainable levels. Around 29% of the world's land surface is under ice, high mountains or otherwise barren, leaving 71% habitable, with half of that used for agriculture. That means that a little over one-third of all dry land has been converted by humans to, essentially, growing food.

However, with drought conditions prevailing over significant areas, climate change, the increasing number and intensity of floods, soil impoverishment, pollution, rising sea levels, agriculture with labour shortages and an increasing number of more demanding customers, suggests a difficult but not impossible future. Ritchie (2024) has pointed out some solutions, if heed to be paid.

1.6 SUSTAINABLE WORLD ENVIRONMENT

Scientists and others, who are contributing to today's climate change studies, must be heeded in order to calculate the carbon footprint and set reduction targets. Fortunately, there are establishments that have been and are studying the problems that derive, but more work has to be done and in varying environments. The findings must be transmitted widely so that the general population, right across the world, must be made aware of the issues that need to be addressed. There has to be a steady drop in carbon emissions, apart from that due to fossil fuels, and with waste and pollution.

Obviously, professionals covering numerous disciplines are trained to play a significant role in implementing these climate mitigation plans. The lack of awareness is worrying which must now be addressed by all countries, international bodies and grassroot communities. Solutions to the solving of problems of hunger and poverty must be found if all life on the environment in which we live and breathe are to be sustainable, according to the opinions of 97% of the world's scientists.

The UN is quoted (in 2022) as saying that, based on a global view, which it admits needs refinement, it is perhaps already too late to 'save' the climate as we know it. Others disagree that that is too pessimistic (CLINTEL, 2022), and now, most likely, we will have to talk about climate change in terms of 'mitigation' or 'adaptation' (NASA, Global Climate Change, 2021). How these now modest targets are going to be reached is the 'Multi-Trillion Dollar Question'. Ambitions to 'bringing nations together', 'putting aside our differences' and 'let's talk it out' at the next COP meeting have limited success.

The solutions are known, backed by sound, conclusive research, extensive testing and a growingly aware public.

In the analyses, the world will be looked at in 32 areas, according to location, topography, environment and climate, rather than viewed on a global average, cross-sectional basis (Koppen-Geiger, 1976; Younger et al., 2022). Moving from the seemingly unworkable political route, appreciating that there cannot be the same answers to addressing the matters of climate change and environmental sustainability for every part of the world, the differing approaches needed will be laid out. Short-term planning, often limited to such issues as parliamentary terms, the remaining term of an authoritarian ruler, or maximising wealth-gathering while the going is good, has not and most likely will not get the world to reach Net Zero GHGs by 2050, for instance. What is happening now must be reversed: the solutions will take even centuries, and programmes will be affected by natural disasters, pandemics and extraterrestrial bombardment. All that must be allowed for in forward planning.

Will there be a long-term sustainable environment in the years ahead? Despite the challenges, there must be. But before implementation, the whole matter must be put to, discussed with and implemented to the satisfaction of all stakeholders. This will mean greater cooperation and understanding among the peoples of all nations to deal with the large environmental problems we face. The means and the knowledge are there. The issues of climate change and environmental degradation must be addressed intelligently, and with full awareness of the drastic consequences if they are not.

The boundaries of science must continue to be pushed out and the understanding of how this planet works. Listen to indigenous peoples and take lessons from Nature and understand that while it has most of the solutions to living on this planet; learning takes time. Short-term, politically aimed, vested interest environmental sticking plaster solutions will not work.

Whether, by the end of this century, there will be human immigrants on the Moon, or indeed on any of our neighbouring planets, looking back to Earth, we do not know at this juncture. Life on Earth has at times led a fragile existence: there having been at least five major life extinction events in the past 600 million years. One of them, at the end of the Permian Period (some 250 million years) wiped out 90% of the Earth's species at that time (Benton, 2003). Non-human animals have been the 'guardians' of this Earth's biome for millions of years. It is up to us now.

The governments that run the different nations, over 200 of them, have been discussed, acknowledging that nowadays two main forms of governance, democracy (in the west) and authoritarianism (Russia and China), have emerged, albeit with variations, from the end of WWII. There are a number of entities, some of which are formerly colonies or individual states, which 'belong' to one or other of the ideas presented by the two forms.

There are some other countries, however, with religion as, the basis of their constitutions, the Islamic countries of the Middle East, for example and others which look back on a historical past with an even older religion, for example, India (Hinduism). In more recent times, a trading bloc of countries called BRICS – Brazil, Russia, India, China, South Africa, – was formed, some of them with the objective of vying with the USA. Recently, the bloc widened its membership mostly with countries around the Middle East. The size of the bloc constitutes 40% of the world's population. It is too soon to say if it will be a bloc for future good governance.

However, it is worth reflecting on what Runciman (2018) says about democracy in his thoughtful book *How democracy ends*. He has observed notable recent changes in behaviour, compared with those given by writers of recent centuries, of political leaders and legislatures, and notes the considerable technocratic changes which have taken place over the past few decades. The interconnectivity today and the developments taking place in AI mean that democracy has also been impacted and will inevitably change. In the end, it just fades away, perhaps not ending but not recognisable as it has been.

This book starts with a discussion on governance and is followed by a contribution on curbing corruption and how best to deal with it. What is written is based on many years of experience in advising global agencies and governments at various levels. Two of the important issues facing government, and the population at large, are then covered, namely education and poverty. A profound discussion is then given on the changing climate and the geological conditions of the landscape, highlighting the value of the surface layers. This then leads to a comprehensive coverage of the environment, rural and urban, rivers and pollution, oceans, etc. which ends the first part. The second part of the book concerns the essential elements which are needed for life and living, namely energy, water and to an extent, transport. Civil Engineers are most prominent in their delivery and consequently the subject is naturally covered with their experience.

The subject of sustainable development is complex, as it has hoped that the book shows with discussion on the main areas that are considered under the broad headings of SEEP issues. With the advent of the Digital Age and with the fairly recent application of AI, it seems that the human experience is accelerating. It would be brave to predict on development beyond one or two decades never mind the end of the century. It is exciting, however, to witness the research and its application in science, medicine and technology and ponder where it will lead. Thoughts are becoming more and more realistic of adventures into space, the Moon and beyond. However, we must not forget the many, many problems that remain on our ancestral home, Earth.

REFERENCES

Ader, C. (1906) The trials of the Eole. In *The Pioneers of Flight*, Scott P. ed. Princeton University Press, pp. *76*–77.

Amnesty International(2025). Women's Rights. Available at https://www.amnesty.org/en/what-we-do/discrimination/womens-rights/

Armstrong, R. (1975) *Powered Ships. The Beginnings*. Ernest Benn Ltd., London.

Benton, M. J. (2003) *When Life Nearly Died. The Greatest Mass Extinction of All Time*. Thames & Hudson, London.

Brown, G. (2021) *Seven Ways to Change the World*. Simon and Schuster, London.

Climate Intelligence, CLINTEL (2022). *The World Climate Declaration*. Available at https://clintel.org/world-climate-declaration/

Copernicus (2023) *The Hottest Year on Record, with Global Temperatures Close to the 1.5°C Limit*. Available at: https://climate.copernicus.eu/copernicus-2023-hottest-year-record#:~:text=2023%20is%20confirmed%20as%20the,month%20in%20the%20ERA5%20dataset

Cook, J. R. and Petts, R. C. (2023). Chapter 1. Roads, development and related policy. In *Rural Road Engineering in Developing Countries*. CRC Press, Turner and Francis, pp. *1*–12.

Dasgupta S., Lall S. and Wheeler D. (2022) *Cutting Global Carbon Emissions: Where Do Cities Stand?* World Bank Blogs, World Bank Group, Washington. Available at https://blogs.worldbank.org/en/sustainablecities/cutting-global-carbon-emissions-where-do-cities-stand.

Friedman, G. (2009) *The Next 100 Years*. Doubleday, New York.

GEM (2022) *Global Entrepreneurship Monitor UK Report 2021/22*. Available at: https://www.enterpriseresearch.ac.uk/wp-content/uploads/2022/09/98522-GEM-Report-UK-2022.pdf

Gross, S. (2020) *Why Are Fossil Fuels So Hard to Quit*? Brookings Education. Available at: https://www.brookings.edu/articles/why-are-fossil-fuels-so-hard-to-quit/

Guardian (2023) SpaceX's Starship reaches space for first time but explodes moments later. The Guardian 18.11.2023. Available at: https://www.theguardian.com/science/2023/nov/18/elon-musk-spacex-loses-contact-with-starship-rocket-after-second-launch-attempt

IEA (2023) *World Energy Outlook 2023*. IEA, Paris. Available at: https://www.iea.org/reports/world-energy-outlook-2023

IPCC (2014) *Fifth Assessment Report (AR5)*. Synthesis Report: Climate Change 2014. Available at: https://www.ipcc.ch/report/ar5/syr/

James, L. (2006) *A History of the Middle Class*. Little Brown, London.

Kirwan, C. G. and Zhiyong, F. (2024) *Smart Cities and Artificial Intelligence*. Elsevier, Amsterdam.

Koppen, W. (1936) Das geographische system der climate (In German). In *Handbuch der Klimatologie* (Koppen W and Geiger G., eds). Gerbruder Bontraeger, Berlin, Germany.
Kremmyda, G. (2021) *Humanitarian Civil Engineering. Practical Solutions for an Interdisciplinary Approach.* Thomas Telford Ltd. London.
Markinblog (2025) *How Many Entrepreneurs Are There*? Available at: https://www.markinblog.com/how-many-entrepreneurs/
Marr, A. (2012) *A History of the World.* Pan MacMillan, London.
Milman, O. (2023) *Global Heating is Accelerating, Warns Scientist Who Sounded Climate Alarm in the 80s.* The Guardian: 02.11.2023. Available at: https://www.theguardian.com/environment/2023/nov/02/heating-faster-climate-change-greenhouse-james-hansen
NASA (2021) *Climate Change.* Available at: https://science.nasa.gov/climate-change/
NASA (2023) *Why Go to Space?* Available at: https://www.nasa.gov/humans-in-space/why-go-to-space/
Niranjan, A. (2023) *Global carbon emissions at fossil fuels to hit record high.* The Guardian 05.12.2023. Available at: https://www.theguardian.com/environment/2023/dec/05/global-carbon-emissions-fossil-fuels-record
Our World in Data (2022) *Poverty.* Available at: https://ourworldindata.org/poverty#:~:text=(2022).,%242.15%20(in%202017%20prices)
Paxman, J. (2021) *Black Gold.* Harper Collins, New York.
Ritchie, H. (2024) *Not the End of the World.* Chatto & Windus, London.
Rothermund, D. (2011) The self-consciousness of post-imperial nations. *Indian Quarterly* 67(1):*1*–18.
Runciman, D. (2018) *How Democracy Ends.* Profile Books Ltd., London.
Rutherford, D. (2013) Living, as it were, in the Stone Age. *JSTOR.* 95:*1*–7.
Smith, A. (1776) *The Wealth of Nations.* Republished 2014 by CreateSpace Independent Publishing Platform.
Statista (2022) *World Population.* Available at: https://www.statista.com/topics/776/population/#topicOverview
Swaddle Team (2023) *The Environmental Costs of War.* The SWDL. Available at. https://www.theswaddle.com/the-environmental-costs-of-war
UN (2018) *68% of the World Population Projected to Live in Urban Areas by 2050.* Available at: https://esa.un.org/unpd/wup/UN (2019) *How Certain Are the United Nations Global Population Projections? Population Facts.* Report. 2019/6. Available at: https://www.un.org/en/development/desa/population/publications/pdf/popfacts/PopFacts_2019-6.pdf
UN (2025) *Sustainable Development Goals (2015–2030).* UN Publications. Available at: https://sdgs.un.org/goals
UN News (2022) *Climate Change: No 'Credible Pathway' to 1.5C Limit, UNEP Warns.* United Nations, New York. Available at: https://news.un.org/en/story/2022/10/1129912
Unger, R. W. (1994) *Coggs, Caravels and Galleons, Sailing Ships 1000–1650.* History of the Sailing Ship Series. Conway Maritime Press, London.
Webster, D. (2009) Regimes in motion: The Kennedy administration and Indonesia's new frontier, 1960–1962. *Diplomatic History* 33(1):*95*–123. Oxford University Press.
Wikipedia (2025a) *The Information Age.* Available at: https://en.wikipedia.org/wiki/Information_Age
Wikipedia (2025b) *History of Colonialism.* Available at: https://en.wikipedia.org/wiki/History_of_colonialism#:~:text=Colonialism%20in%20the%20modern%20sense,would%20soon%20reverse%20and%20surpass
Younger, J. S. (2007) *Sustainable Development.* Guest lecture to Glasgow University Sustainable Development Network.
Younger, J. S., Parry, D. E. and Meigh, J. D. (2022) Preparing for the future: The impact of climate change on the civil engineering profession. *ICE Proceedings Civil Engineering* 175(2):*1*–8.

Government and Corruption Issues

2

Scott Younger and Hamish Goldie-Scot

2.1 INTRODUCTION

This chapter is initially concerned with the forms of government that are found today and what it involves for the population. For authoritarian governments, there is little or no participation and therefore the people effectively have no say in the way that its nation is run. Countries that have this form of government are those that have taken up the communist message, albeit somewhat modified nowadays, Russia and China being the most prominent ones. At the other end of the scale, there are the democracies, the nations with a western view, the US, UK, the European nations mostly, Japan, India, South Korea and British commonwealth nations, for instance, where the population is involved to a greater or lesser extent. These involve layers of governance of different complexity, where the people can be directly involved. There is also quite a number which are in between, as well as theocratic ones, some extreme, with women facing severe strictures. Altogether, the population is spread among about 200+ nations.

Secondly, no country, except perhaps the very small, escapes the problem of corruption which varies from the relatively trivial to the very significant. Much depends on how authoritarian the government is and what is the basis of that government's power base, e.g. drugs as in some South American countries. Acknowledging that even democracies are not free of corruption, the thinking is to find a reasonable way of curbing it. The Infrastructure Transparency Initiative (CoST), for example, is presented as a system which has been developed to curb corruption on infrastructure projects and thereby improve the return on the investment made and hence an aid to Governments and Governance.

 DOI: 10.1201/9781003566274-2

2.2 GOVERNMENTS AND GOVERNANCE

2.2.1 Historic Context

As societies have developed over the ages, communities have realised that there were certain communal decisions, which had to be taken by those who generally showed appropriate leadership qualities. As the communities grew in size, so the early forms of self-governance became bigger until, as of today, every country has some form of government, usually divided into different strata, each of greater significance than the level below. Unfortunately, sometimes the wrong leadership occurs or grabs the power, and life becomes difficult for the citizens, since that power becomes dictatorial in time, if not from the start.

Today, there are over 200 entities, 196 are members of the United Nations (UN), which reflect past movements of peoples, ethnic traits and the quirks of history. While adjustments take place among different communities of a country, with modern communications, sometimes instantaneous, it means that generally these adjustments will be small, probably only involving physical boundaries. Occasionally, through an act of war, a country is unwillingly taken over by another. The most significant change taking place, over the past few decades of last century until today, is in the growth of the world population, as shown in Figure 1.1, and in the expansion of urban communities. In this period, the population expansion was largely in Asia, China and India. For the past decade and the immediate years ahead, it will be that of Central, Sub-Saharan Africa. A major catastrophe taking place, although rare however, may cause some country mergers as necessary for survival in the years ahead.

For countries that include several million people, the government is drawn up to carry out many functions, generally called public services, which are meant to deal with the requirements of citizens of the country. Furthermore, the government is generally given the power by the citizens to handle such issues as national domestic matters, and also to deal with foreign and trade relations. As government assumes the management, it requires payment for its undertakings, and this payment is garnered by charging the citizens through taxation. Theoretically, no government has funds of its own and raises money from the people on behalf of the people and allocates it according to a consensus of the people, usually debated in a chamber of representatives elected by the people. This system is known as a form of democracy, in which the individual has considerable personal freedoms, but the amount of personal freedom varies from country to country and large strictures are placed in some countries whereby the form of government veers towards dictatorship. The greatest amount of personal freedom is shown among western powers, e.g. Europe, US, India and other UK Commonwealth countries, most of ASEAN, and countries on the east Asian fringe, Japan, South Korea and Taiwan.

Compared with just two centuries ago, the number of citizens in many countries nowadays is so large that a national chamber or parliament cannot cope with all matters, nor would it be sensible, and consequently different levels of localised government are needed. These vary from country to country, except for a few exceptions, but can be all or some of the following: state, county or province, then district, cities, towns, villages and ultimately small communities, like hamlets.

Today, some cities have become so large, particularly in the past century, that they are known as megacities, which have about 10+ million inhabitants. One such city is Tokyo which has greater than 30 million within its greater metropolitan area and has earned the sobriquet of meta-city. Delhi in India has also passed the 30 million mark. The expanding of cities is continuing in several countries. By the 2030s, the cities of Jakarta and Bandung in Indonesia are expected to have effectively merged into a country-size unit of approximately 70 million people. The biggest expansions of populations have taken place in South Asia (India), Southeast and East Asia (China), along with some countries of Sub-Saharan Africa, such as Nigeria. Uganda's population has increased by a factor of approximately 7 since 1960 to stand at 47 million today. The continent of Africa is forecast to have a population of 4.2 billion by the end of this century, which will be the fastest area of expansion in the world this century. It will, by then, be approximately 37% of the world population. The other notable change, as discussed above, will be China, which is already declining from a peak of over 1.4 billion to be approximately 800 million by 2100 as a result of an austere one child policy for two decades in the late 20th century, leaving a demographic imbalance.

However, in today's world there are broadly speaking two main types of government operating, the first being a democratic form and the second authoritarian or even dictatorial where there is only one level of government. There is also a third, theocratic form of government, which is based on the somewhat strict rigid beliefs in some forms of Islam, in particular. These theocratic forms are found in certain countries of the Middle East and then extending to southwest Asia and northeast Africa. Their governments, arguably, can be considered as unsustainable in the long run because they are too rigid and too restrictive. Examples today are Iran and especially the extreme Taliban government of Afghanistan.

Two other countries which have restrictive, and, in the long run, unsustainable forms of government are North Korea, which has a somewhat delusional leader, who spends most of his country's funds on dangerous nuclear weapons, and Myanmar. This country, Myanmar, has been taken over by a small military elite in a coup and the democratically elected civilian government has been quashed. War crimes are being committed regularly, and resistance is building, almost civil war conditions. So far, the free democratic countries have not found a way to overcome this unsatisfactory situation without interference, and watch and contain the influence, unfortunately with little success. Some Sub-Saharan countries have unstable dictatorial governments, mainly arising from poverty and/or corruption, and similarly some in South America, but these latter often include powerful families with large dealings in illegal drugs.

The autocratic governments of Russia and China have been the outcome of each espousing the principles of communism a little over one century ago in the case of Russia, and after World War II (WWII) in the case of China following a hard-fought civil war, In fact, each of them, Russia and China, had an out-of-date dynastic structure, the likes of which other European countries had already rejected or left with limited powers, albeit comparatively recently in some cases, and adopted a democratic form of government. Russia had had a Tsar until 1917 and China an Emperor, each of whom ruled with supreme, and supposedly inviolate powers, a form of autocracy. This then was toppled in each case and, after an unsettled period and civil wars, saw communist elements appearing as a small but powerful elite group, following the 19th century teachings of Karl Marx (Marx and Engels, 1848). Communism took charge and prevailed in each of those two countries to this day, although some components of capitalism have crept in and are part of business dealings and trading.

The average country across the world, say with 5–10 million citizens, has four or five jurisdictions, each with a defined level of responsibility, and tax raising ability, which is limited to districts and towns, to a defined extent, but not to lower levels. That is, to local government, where there are officials who are fully employed to raise and handle funds not requiring central control and approval but answerable to a number of selected citizens on behalf of the rest of the people by whom they are elected.

In much of the early discussion that has occurred concerning sustainable development, since the term appeared in the late 1980s/early 1990s, the emphasis was on social, environmental and economic issues. Political ones and, by implication, legal were not considered. This was commented on by Younger et al. (2012). It has been found that, in nearly all cases, the political issue cannot be ignored if one must make a comprehensive judgement of the development of a country. This draws the distinction with growth which largely is a measure of how well a country is deemed to be doing in economic terms.

One cannot ignore the human growth impacts on the world. Earlier, a significant increase in the population in the past 60 years was discussed. The influence that this was having on the increasingly unsettled human population and the natural world has not yet been fully comprehended. The governments of today, sometimes through the 77-year-old institution of the UN, to which 196 countries except a handful of nations of the world effectively belong, are concerned with the apparent deleterious effect that humans are having on the planet. This can be largely attributed to the fast growth in population within a short period and thereby a significantly increased call upon resources, e.g. water and energy, and also causing of more waste and pollution.

The world is our natural home, at least for the foreseeable future, although there is a slowly growing awareness of the possibilities that other nearby planets may offer. For the vast majority, however, there is also a realisation that much more has to be done on Earth in terms of its care and protection. Had the population expansion over the past 60 years been on a trajectory like the previous period, the population today would be closer to under 5 billion and very probably the extent of the climate-related and other problems would be less marked as would the pressures that the population numbers themselves have evoked.

The UN have drawn attention to what is termed "climate change", and the apparent heating of the Earth which, if not contained, will spell very difficult, unsustainable conditions for life. However, there is a view that this is too simple, and other factors must be considered, as discussed subsequently. There are a number of governments that do not follow the UN's guidance, e.g. limiting greenhouse gases. Sometimes it is because it is seen as contrary to a country's own plans for development through economic growth and defined priorities, but a lot through just plain ignorance and lack of funding to undertake the actions to curb climate change that the UN deems and some recent events would indicate as necessary.

There are major issues with pollution, for instance, which at times transcend a nation's boundaries, and require international legal intervention, particularly for issues that involve the oceans, such as raw sewage discharge, usually via rivers, and the dumping into them of non-degradable plastics. The pollution levels from their use have grown at an unappreciated and unprecedented fast rate in the past 60 years, both on sea and land, in line with the population expansion and have a negative effect on the environment. It has been estimated that the annual production of non-degradable plastics has reached about 400 million tons of which only about 10% is subject to treatment. This involves a major health hazard, the extent of which is not yet known, although work is coming to fruition on treatment of plastic waste (Neary, 2026).

There is also an increasing problem of sewage, which has only now been coming to attention. Facilities have not expanded with rate of increase of population and many plants are old and not fit for purpose. Many of these treatment works are significantly under capacity and ultimately risk discharge of untreated, or partially treated, sewage into either a river or the sea.

Poverty is an uncomfortable matter for all governments, or should be, but particularly for poor countries with a sizeable population. With the percentage of poor in a growing population barely improving with the passage of time, the total number of people in poverty has increased in total. It is thought that the total number of people in the world today so affected is well over 1.4 billion, which means that the people in poverty have more than doubled in half a century, although as a percentage of total population, it remains the same. Most people do not want to live in poverty but do not know how or have the ability to escape its tragic, debilitating consequences. It is a highly significant problem for many governments, particularly so in some countries of Africa and Asia, not to mention those who receive the thousands of migrants who leave their poverty-stricken homes seeking a better, more secure future. The UN, one of whose mandates is to reduce if not eliminate poverty, as expressed in their Sustainable Development Goals (2015–2030), tries to assist in some of the poorest countries, but their resources, in terms of application of manpower as well as finance, are limited.

Human life expectancy has doubled over the past century from 36 years in 1920 to more than 72 years today (Lomborg, 2022). A hundred years ago, three quarters of the world's population lived in poverty. Today, it is less than one-tenth, although the overall number in poverty is similar to that before because of the great increase in overall numbers. Whatever there is marked improvement in the life of much of humankind, but still too much poverty exists.

Much of this significant improvement in longevity can be attributed to medical research and practice and improvements in agricultural output, not to forget about education which has greatly increased in importance in some, perhaps most governments' agenda. In turn, people are living longer with the average age of an individual getting noticeably older and more and more dependence being put on the younger generation. Society is changing, most probably faster than anytime previously, but there are significant demographic differences between older, more mature countries, as found in Japan or Europe or the US, and the younger countries, as located in Sub-Saharan Africa. There are therefore different challenges that governments have to meet.

Altogether, this will mean gradual changes in the ways that we organise society and, in turn, how it will be best served and governed. The forecast is of slowing growth henceforward in the population as mid-century is approached, after the recent spurt, and thereafter a gradual decline from about 2080 to a level of approximately 10.9 billion at the end of the century (UN, 2025). This is thought to be an acceptable level which can, it is believed, be coped with, and would suggest that a sustainable level of population can be reached over this century. The usual checks to sustainable progress, both natural and manmade, will be around. The natural, for example, weather-related and internal earth issues we have to learn to cope with, and the man-made we try to resolve through diplomacy, where we can! The evidence of the last couple of centuries would suggest that there is a long way to go to reduce occasions for squabble and worse, war and strife. They can be expected to be still worryingly unresolved issues throughout this century, with tools of mass destruction readily to hand. In terms of method of governance, the world has changed little in the past decades, but it is worthwhile to examine in some detail the happenings of the past seven decades.

Friedmann referred to the 500 years until the mid-20th century as the European Age (Friedmann, 2009) and added that the 21st century would be most influenced by the US, this dominance declining towards the latter stages as the 21st century approached. The forthcoming 60–70 years from today suggest more scrutiny, in particular following more recent, the last decades, changes in society, as voiced in Runciman's timely book (Runciman, 2018), *How Democracy Ends.* There is also the development of artificial intelligence (AI) which is coming on at a rapid pace. Governance of communities, not just nations, will have had to adapt to significant changes in how people will live.

2.2.2 A Brief History of More Recent Events

A key feature of the European 500 years was colonialism which, more or less, drew to a close a decade or two after WWII; it is generally agreed that it was an imposed injustice by certain countries of the world. This view was expressed through the new voice of the UN. It is pertinent to be reminded at the growth of Asia over this period, post-1950, as highlighted in Figure 1.2.

In the 1950s, North and South Korea clashed, each side bringing in opposing forces, the north siding with the communist regimes of Russia and China and the south with western democracies led by the Americans. No outright victor emerged, and Korea was divided between north and south along the 38th Parallel, as it remains to this day. North Korea's economy, nominally based on the communist philosophy, has not advanced in the intervening years and the North Korean people are very hard pressed by their government with its "unusual" leader. On the opposite side, South Korea has blossomed and can be described as a developed country along with Japan and to an extent Taiwan, ROC, because the latter's independent existence is disputed by China. The three, South Korea, Japan and Taiwan make up the east Asian fringe and rank among the developed nations.

In the 25 years or so following WWII, US foreign policy in Asia tried, often unsuccessfully, to balance two sometimes contradictory aims: firstly, stop the spread of communism (the domino policy) and secondly, discourage the continuation of colonial rule by its European allies. In Vietnam, Laos and Cambodia, this mixed message policy had disastrous consequences. While supplying the French with arms and materiel post-1946 to fight communism in Indo-China, the US initially stopped short of getting involved in fighting. However, after the French defeat at Dien Ben Phu (in 1953) and the final departure of the last French troops (in 1956), the fear of communist takeover led the USA to get gradually sucked into disastrous air and ground wars from around 1964 onwards. The eventual US withdrawal allowed a reunification of a nominal communist Vietnam (in 1974) as well as the emergence of independent and communist-dominated Laos and Cambodia (Tucker, 1999). Although Vietnam is now recognised as a Middle-Income Country, the development and social consequences of these events are still being felt today, particularly in Laos and Cambodia.

By the time the 1970s arrived, if not before, America realised that they were looked upon as leaders of the "western" world. The long-standing colonial powers in Asia – Britain, The Netherlands, France and to a small extent, Spain and Portugal – were much diminished, and their world influence was reducing noticeably by the decade. New countries had appeared on the scene or old countries rebadged – Indonesia (formerly the Dutch East Indies), the Philippines (historically Spanish or USA dominated), Thailand (a kingdom, formerly known as Siam, uncolonised), Malaysia (comprising British peninsula Malaya, put together with Sarawak and Sabah on the island of Borneo), the Sultanate of Brunei (independent but with strong British connections), Vietnam, Cambodia and Laos (all three formerly making up French Indochina), and the island of Singapore, a Chinese conclave breaking off from the peninsula of Malaysia in 1963. Nine countries, including Burma (later Myanmar), became the Association of Southeast Asian Nations or ASEAN, as it is known today, a useful trading bloc, recognised by the EU, US and UN, and very recently, China, following the imposition of trade tariffs by the US government. The original grouping, the founders, of ASEAN comprised just five countries, Indonesia, Thailand, Malaysia, Singapore and the Philippines.

Before this original group of countries was brought together in Bangkok in 1967 as ASEAN, President Soekarno, of Indonesia, had sought to interest a host of non-aligned countries from different continents. A body was formed in Bandung, Indonesia in 1955, PM Jawaharlal Nehru of India taking a prominent role. Nehru and Soekarno had known each other previously in restless Europe in the 1930s.

Both the non-aligned movement and ASEAN are based in Indonesia which confirms the seriousness which Indonesia takes the importance of these bodies in terms of the economy of the region, trade and peace. The importance of the non-aligned bloc is that the 42nd Summit in recognition of President Soekarno's role in starting the bloc, was held in Jakarta in September 2023, and opened by the President of Indonesia, Joko Widodo. There were strong sessions also with the UN, the US and India, and closure was chaired by the Philippines.

Alongside ASEAN, there is a grouping of South Asian nations with India at its core. The nations are Pakistan, Afghanistan, Nepal, Bhutan, Bangladesh, India, Sri Lanka, British Indian Ocean Territories, Maldives, and now and again Iran. India, from its areal size and very large population, dominates this bloc of nations. This grouping of nations is not so far advanced structurally as ASEAN.

As 1980 approached, nearly all parts of Asia were steadily growing in terms of their respective economies as well as populations, following Japan and South Korea and advanced countries like the US and China, which would in due course overtake Japan as the world's second largest economy in terms of GDP, the countries that made up ASEAN, and the South Asian contingent centred round India. This was marked by the impressive growth of existing cities and the burgeoning roots of new towns and cities in most countries, and the development of subnational governance.

The last two decades of the 20th century saw a marked change in the world. The Soviet Union, USSR, which had assumed sovereignty over East Europe after WWII, including East Germany, the boundary of which with the western nations passed through Berlin, was crumbling; it was no longer going to be able to keep up economically with America. It had tried the communist approach to government for nearly four decades and found it was inefficient, led to food shortages, made poor use of manpower, and it was falling further and further behind the US. It could no longer control the countries of Eastern Europe, which they had held within the "Iron Curtain", since WWII, although their grasp on Eastern Europe took several years to unwind. The 1980s was the time of new Russian leaders, e.g. Mikhail Gorbachev, who paraded the ideas of *glasnost* and *perestroika*. But the most dramatic event took place in November 1989 with the Fall of the Berlin Wall. This was put up in 1961, because the Soviet Union became alarmed at the number of East Germans finding their way across the border with the west to seek better opportunities. It still took a few years before the countries of Eastern Europe to gain their independence from the USSR and express their own identities. But trouble was looming in the Balkans; Yugoslavia was on the cusp of breaking up.

In the 1990s, as these developed, the countries of the Balkans, disintegrated further into their constituent parts, Serbia, Bosnia, Croatia, Montenegro and Slovenia, basically along religious lines and where the Ottoman empire had held sway, until the early years of the 19th century. They had been held together as Yugoslavia under a strongman, Marshall Tito who had emerged during WWII, supporting the western allies and formed the southern boundary of the communist-held lands in Europe. When Tito died in 1980, the Soviet Union was nearing its "death" throws and the breakup of the Balkans from the constituent parts was inevitable. The resultant strife was nasty, and peace was restored by western countries fighting under the UN banner, separating the warring factions. However, the nations of the Balkans have generally today settled down to their constituent countries.

In the 1990s also, Al Queda, an Islamic terrorist organisation, began to make its presence felt. Further, there was the possibility of getting an accord between the Israelis and the Palestinians, brokered by the US and western allies, but extremists on both sides prevented this happening. It remains an unresolved issue to this day. The new century saw the US becoming involved in the Middle East, the countries of Iraq, Iran and Syria, in particular. Strict, sometimes extreme, forms of Islam, on top of the Al Qaeda group, were promoted by Iran and a body called Islamic State. However, the attention of the US was undoubtedly held in 2001 when Al Qaeda terrorism reached their shores, and the 90-storey twin towers in New York were destroyed utterly, approximately 3,000 people killed, in an infamous terrorist attack on 11th September.

A few days later George W. Bush, in his second year as President, declared war on Afghanistan, reputedly Al Qaeda's base of operations, and the US fought ostensibly an unsatisfactory campaign until they withdrew. Afghanistan remains a difficult and elusive country to fight in, as the British and Russians have found in the past and would agree. The extreme Islamic fundamentalists, named the Taliban, have worked their way back and currently hold the reins of government and suppress the people, particularly the women.

A recent addition to political-economic groupings of nations is the BRICS bloc, from an initial idea of Russia in 2006. The bloc, which encompasses Brazil, Russia, India, China and South Africa, was formed in 2010 (initially without South Africa who only confirmed their participation in 2009). The bloc includes about 40% of the world's population and excludes Europe and the US. It includes the two still espousing the nominally communist way the other three from the non-aligned movement. Its most recent summit meeting was in S. Africa in September 2023. More recently again, 1 January 2024, the bloc widened its membership with the following countries joining – Saudi Arabia, UAE, Egypt, Iran and Ethiopia. This, in a way, expresses a profound disagreement or dismay at the support given to Israel by the US in the current Israel–Palestine conflict and gives some heart to Russia in its wrongful attempt to take over Ukraine.

An overview of the years since WWII shows a complete change from that time, and even the forecasts of Friedman (2009) for the century ahead will likely have to be altered, as he himself also stated. The forecasts for the century ahead did not take account of BRICS, and its recent extension of membership, and the current strife in Ukraine, which are signs that the current regimes of China and Russia would appear to have halted their trends of opening up to democracy, which they seemed to be doing at the turn of the century. To this must be added the more recent addition of AI as introduced above and the tremendous impacts on daily life.

It is timely to look at how the US fared over this period since WWII with an intent to see how they might do over the rest of this century. Their scientists have also developed a significant number of technologies over this period and must be aware that they have promoted a good many of the cutting-edge discoveries. At the same time, particularly in the early decades after WWII, there was a great fear of the spread of communism, especially where the country had a recent colonial past or was otherwise undeveloped. It affected the US thinking which, in turn, undoubtedly coloured its attitudes towards other nations throughout the 21st century, and theirs towards the US.

Friedmann (2009), in his overview of the rest of this century, forecast that there would be change as the century unfolded, with the US waning in importance in the latter stages of the century. He chose Poland, Turkey and Mexico as being countries of substance to watch. No doubt he would revise his choices as the decades wore on. Even today, the world looks different from what it did 20 years ago.

2.2.3 Governance

There must be continual attention to governance. The history of this is distinctly variable and would seem to depend on the leaders within the system adopted by or foisted on or dictated to the people for a given situation, the length of time the system that has been in place, and whether there is a religion which has a control on the way of life. The UN is a global body to which almost all governments belong, as said, some taking membership seriously and others paying token lip service, usually, but not always, the most undeveloped countries, sometimes through ignorance. Perhaps, in a given country, more important notice should be taken of decision-making at the level of a town to improve services.

The pace at which knowledge has become available has accelerated impressively in this Digital Age with continuing advances to the computer and computation. No longer is rote learning, as was the system of teaching school children 50 years ago and until this century, accepted as the only way to teach today. However, there is still a place for it, at least at primary and sometimes elementary level, particularly at the poorer end of society. In advanced schools, they are already experimenting with AI. It is growing in importance as more and more individuals and corporations, especially in the more developed countries, are experimenting with it, and seeing how it can be used, for instance to replace mundane jobs in labour. It is difficult to predict where AI will lead, and there are bad uses of it which, if it were promoted by the wrong people, could lead to some dangerous outcomes. But it is here to stay, and we must take its many good points and embrace it.

This will affect the way that government works as the century proceeds, but government will keep attracting people who will be behind the younger generation coming up. This is a factor becoming more and more discernible nowadays but likely to become wider. However, education will have to cater for an even bigger range of abilities and levels of opportunity starting with those unable to avoid poverty and caught in its trap of little chance of even basic learning. Early life education is absolutely vital for nurturing a healthy society.

The Mother of Parliaments, the oldest democratic chamber, as the UK House of Commons is known, is showing its age, and suggests that some of its procedures should be overhauled. The upper chamber, the House of Lords will possibly be changed in the next decade, and become an upper chamber where members are elected on merit or popular vote and not because of centuries-old heredity values or a "thank you" for political party donations or services rendered. The US also have two chambers, the upper called the Senate and the lower the Congress, the House of Representatives, but both are elected. Major change, however, will be slow. The EU is trying to find equilibrium among disparate European nations with quite different histories, some countries even having been at strife with each other in the fairly recent past and still hold grudges, which occasionally boil over to violence. Will autocracies, dictatorships survive? Probably some, a question of time, perhaps a few generations, depending on the will of the people so affected. Will democracy survive? Probably but in fading manner as discussed above.

What will the styles of government be like in the medium term? Probably not much different from today, but it is interesting to speculate that the difference between the most advanced compared with the poorest will widen. Estonia, for instance, with only approximately 1.4 million people has a young-in-age government and pushing cutting-edge ideas, as with AI. They are attracting an immigration of significant numbers of university-educated people, Estonian and otherwise, with the aim of becoming one of the leading countries of the world, and the same will spill over into adjoining countries or at least to neighbours and the Scandinavian countries to the north. However, there is still too large a number of countries that have significant relatively uneducated populations, which unfortunately will not show much progress, although there will be a greater number of people altogether in total educated terms.

Education – and training – is changing today, as discussed more fully below; there is talk, in many outlets of the media, of AI, which is being introduced in some schools, as discussed. More people, except perhaps the over 70–75 generation, are totally media-savvy. By mid-century, those aged over 75 will face different, but similar in effect, challenges to today's older generation. Governments will still be lagging behind the "will of the people".

Before the end of the century, some people, albeit few in number, may well be taking trips to the Moon for a holiday, assuming that the countries involved can reach an accord. At the other end of the scale, we could or should be still worrying about whether we are doing enough regarding the environment and on the issues that affect poverty. The UN Sustainable Development Goals will be into their sixth makeover, assuming if updates every 15 years.

The numbers taking an interest in space by the century's end will still be very small but will stimulate the human interest. They will widen the gap in knowledge, still building on what we know today. But the most important concern will be if the countries that are involved in space exploration, particularly the US, China, Russia, EU, Japan, India and UK, for instance, will cooperate and not risk bringing disaster, e.g. an unknown virus, to home base, Mother Earth (Marshall, 2023). There will need to be an order to governance in this newest extension to mankind's dominion.

By the end of the century, the future will look quite different from that today, as daily life shows significant changes to that even experienced some 70 or so years ago.

2.3 CURBING CORRUPTION

2.3.1 Introduction

Any initiative aimed at enhancing sustainability in relation to public infrastructure and other developments runs the risk of a veneer of compliance. Without associated good governance, this can mask poor decision-making, resulting not only in wasted spending, but also in a failure to ensure that the requisite quality is achieved from a technical, social and environmental perspective.

This section provides an historical perspective on why corruption has proved such an intractable problem, and how a tendency to focus unduly on details of the problem rather than on underlying causes and practical solutions has proved unhelpful. Not only does the effort required to study how corruption is manifested provide a false sense of achievement and progress, but also the adoption of unduly superficial responses to what is a highly complex, multi-faceted and deep-rooted issue inevitably gives rise to some unintended negative consequences.

Though there are no easy answers, there is growing evidence of the effectiveness of broader strategic approaches that accept that corruption risks will always exist, but which seek to manage the underlying drivers of those risks in a more coherent, constructive and collaborative manner. This includes focusing on the shared interests of all stakeholders, not just clients and providers of goods, services and works, but also investors and intended beneficiaries. Albeit in different ways and to varying degrees, each has an interest in the achievement of value for money in the provision of infrastructure, goods and related services that meet citizens' needs while contributing to climate-related goals.

2.3.2 Framing the Problem

Though there is no universally accepted definition of what constitutes corruption, it is generally considered to entail the abuse of entrusted power for private gain (Transparency International, 2023).

Corruption can be considered from various perspectives. In terms of illicit financial flows, it is estimated (UNODC, 2023) that the amount of money laundered globally in 1 year is between 2% and 5% of global GDP, or about $800 billion to $2 trillion in 2022. Fines handed out for global financial crimes have typically totalled between about $10 and $22 billion annually (Napier, 2022).

From a sectoral perspective, perception surveys typically identify three sectors particularly prone to corruption as being the extractive industries, defence and construction. [A Global Economic Crime Survey conducted found the engineering and construction sector to have the highest rate of bribery and corruption of any industry (PwC, 2014).] Compared with other sectors, each of these is typically characterised by a combination of very significant spending and relatively weak associated accountability mechanisms. In each case, this can entail a combination of a lack of transparency, limited capacity to make effective use of what data are available, and powerful actors with a vested interest in maintaining associated illicit financial flows. Other sectors such as health and education also suffer from many forms of corruption, including, but not only, those related to the procurement of related infrastructure.

In most countries, infrastructure investment constitutes a significant proportion of public procurement, yet is particularly prone to risks of inefficiency, mismanagement and corruption. A recent International Monetary Fund (IMF) study (IMF, 2020) concludes that such wastage amounts, respectively, to 53%, 34% and 15% of total spending on infrastructure in developing countries, emerging market economies and advanced economies. Recent high-profile convictions in OECD countries for infrastructure-related corruption highlight the fact that shortcomings in sector governance, and associated corruption risks, are not limited to low-income countries but remain a global challenge. Notable examples include successful corruption prosecutions related to Berlin's Brandenburg Airport, which in 2020 was finally completed €4.5 bn over budget and nine years behind schedule (Business Insider, 2020), and Canada's SNC-Lavalin Construction pleading guilty in 2019 to massive fraud charges related to its contracts in Libya (Reuters, 2019).

An important concept that is reflected in the IMF study is the close inter-relationship that exists between corruption and mismanagement, typically in the form of a lack of capacity and/or ineffective accountability mechanisms. This is consistent with the results of research undertaken for the Commonwealth Secretariat described in an online training programme (ComSec, 2021a). As illustrated in Figure 2.1, it is found that, rather than viewing corruption as a cause of poor management, civil servants generally

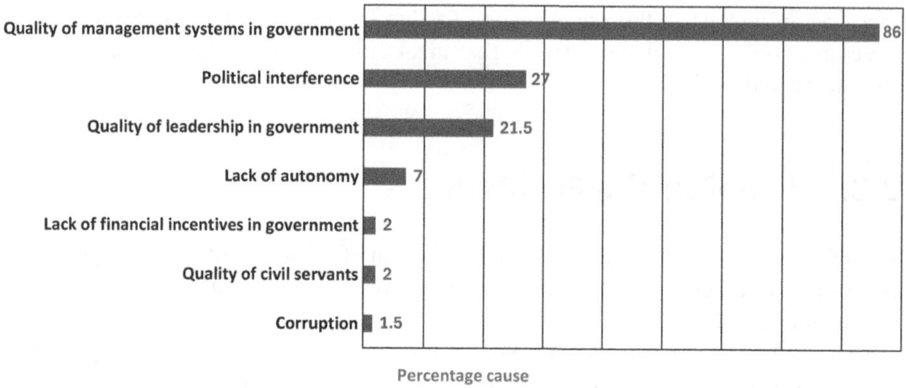

FIGURE 2.1 Perceived causes of poor performance in Commonwealth countries (ComSec, 2021a).

see poor management systems as giving rise to poor performance, while also opening the door to corruption risks.

Taken further, this suggests that corruption might even at times be viewed as a practical solution to an underlying management failing. Either way, a focus on improving management processes should help to ensure that associated corruption risks will be reduced.

Another way of looking at this is that it is helpful to focus on the solution rather than risk simply studying the problem. Specifically, "fighting corruption" should not necessarily be viewed as an end in itself, not least because many people involved in, or turning a blind eye to, corruption consider that they have no choice (Goldie-Scot, 2020). The core problems to be addressed are not simply those associated with corruption itself, but the fact that it undermines good performance and manifests itself in secondary risks. From the perspective of sustainable development, opaque and unaccountable planning and procurement processes result in critical infrastructure being unnecessarily exposed to the risk of climate change.

Even without any corruption *per se*, a simple lack of capacity in effective planning, tender management and delivery can easily result in infrastructure that is not fit for purpose in terms of quality, value for money, sustainability or relevance to the needs of intended beneficiaries. By focusing on solutions to such problems, many associated corruption risks will also be addressed.

In developing this approach, the primary focus will be on corruption related to public infrastructure as seen from a civil engineering perspective. However, the resulting conceptual framework and associated tools for understanding and addressing the challenge can also be helpfully applied to other sectors.

2.3.3 Historical Perspective

The current widespread criminalisation of corruption and adoption of "zero-tolerance" policies in relation to corruption risk management are very recent developments.

Civil engineers are bound by their professional standards to an overriding responsibility to the public good. In the case of members of the UK's Institution of Civil Engineers (ICE), founded in 1818, it remains the case that "Members of the ICE should always be aware of their overriding responsibility to the public good. A member's obligations to the client can never override this…" (ICE, 2022). Yet during the Cold War, this responsibility to the public good was quietly but explicitly interpreted by some Multilateral Development Banks (MDBs) as requiring its economists, engineers and others to turn a blind eye to high levels of corruption on infrastructure projects. The rationale for doing so was to help ensure that recipient countries aligned themselves with the west, rather than with the Soviet Union.

Robert Klitgaard's book *Controlling Corruption* (Klitgaard, 1991) helped raise awareness of the many and varied risks associated with corruption, and the establishment in 1993 of Transparency International provided a solid foundation for research and related advocacy.

Most international institutions were remarkably slow to take the issue seriously. When James Wolfensohn arrived at the World Bank in 1995 he was "not allowed to talk about" the "C-word" (World Bank, 2005) and it was not until his "cancer of corruption" speech in 1996 that the Bank moved towards a less nuanced stance regarding corruption in relation to its operations.

To appreciate just how much has changed, it is helpful to remember that, at that time, paying bribes or facilitation payments to foreign public officials as a way of obtaining or being allowed to compete for international contracts was considered to be a "perfectly normal way of doing business in many OECD countries" (Cole, 2015). In most cases it was even tax deductible.

The US was well ahead of the curve here, with its Foreign Corrupt Practices Act already prohibiting US citizens and entities from bribing foreign government officials to benefit their business interests. From 1977, this had been applied specifically to publicly traded companies, but following amendments made in 1998, it also applies to foreign firms and persons who, either directly or through intermediaries, helped facilitate or carry out corrupt payments in US territory.

At the other end of the scale, the UK was behind the curve, and long criticised by the OECD as being a late starter when it came to legislating against corruption (OECD, 2008). When the UK Bribery Act finally came into force in 2011, it gave rise to a new offence of "failing to prevent corruption". This introduced an associated acceptable defence if it could be shown that "adequate procedures" were in place in the form of policies, procedures, and related training to guard against such risks.

Fast forward 10 years, and "anti-corruption" had itself become big business, with very high levels of spending on anti-corruption research, legislation, enforcement, and training. Yet as the IMF study found, significant underlying problems remain, and corruption risks remain at every stage of the project cycle, and at every stage of the procurement of constituent contracts.

2.3.4 Unintended Consequences

There are many reasons for this lack of progress. First and foremost is the fact that corruption exists in the shadows and can adapt very quickly to changing circumstances.

Add to this the tendency of policymakers to focus on the symptoms of corruption rather than its underlying causes, and of public officials (and MDB staff) not to rock the boat, and it is easy to see how known problems are allowed to persist.

As set out in Box 2.1, a simple anecdote (Goldie-Scot, 2001) from a country in Asia neatly illustrates these three factors.

BOX 2.1 WHACK-A-MOLE: HOW CORRUPTION REINVENTS ITSELF

In an attempt to counter high levels of corruption in the local purchase of minor items of equipment, an MDB developed and introduced strict new rules for local purchasing, requiring quotations to be provided from three different suppliers. The new requirement took years to formulate, but within days of it coming into effect, many local suppliers had secretly agreed to share their stationery, so that any one of them could supply three "quotations", thereby saving everyone both time and effort, and facilitating the usual deals providing significant incentives for the purchaser. When this was informally reported to the MDB's country office, the response was not one of surprise, but rather a weary "I didn't hear that". In time, further safeguards were put in place, and again soon circumvented. This process continues, with ever-increasing sophistication.

Similar cat-and-mouse processes have occurred in the procurement of services and works, as successive safeguards are put in place in response to violation of the rules. Eventually, the rules themselves can ironically become a driver of corruption. Not only do they generate powerful incentives to take shortcuts, but perhaps more importantly they also send a signal to contractors that they are not trusted. Such a lack of trust in the procurement process inevitably gives rise to time delays and increased costs, while contributing to associated corruption risks.

A related example of unintended consequences derives from the trend towards clients preparing ever more prescriptive Terms of Reference in response to instances of poor performance by contracted parties. In the case of procuring supervising consultants on major road projects, this has reached the stage of some MDBs defining exactly how the successful bidder should staff the project and organise their resources. As a result, the difference between a "good" bid and an "excellent" one may only be a matter of a few percentage points, as every bidder seeks to give the appearance of meeting all the specified requirements for both staffing and methodology. Even assuming there is no associated corruption, this can undermine the notion that Quality and Cost-Based Selection helps ensure a focus on quality. A better way of differentiating between competing technical proposals would arguably be to leave it up to bidders to set out how they would fulfil their responsibilities for administrating the contract and supervising the works. For such an approach to succeed would, however, require both willingness and the ability on the part of clients to hold companies to account for their performance in fulfilling contractual commitments. In many cases, either such an ability, or such a willingness, is lacking in practice.

One of the responses of MDBs to the problem of demand-side capacity constraints and associated corruption risks has been the establishment of Project Implementation

Units (PIUs). Intended to serve islands of integrity and a training ground for "best practice" in infrastructure procurement, these can give rise to the development of more sophisticated forms of corruption, while maintaining a veneer of compliance. In terms of sustainability, a further negative impact can be to undermine the capacity and competence of mainstream government (ADB, 2005), while further diminishing any sense of accountability for performance.

Even when a PIU has the necessary capacity and is operating with integrity, private sector capacity constraints can undermine performance. The continued insistence of some MDBs on the use of adversarial forms of contract based on priced bills of quantity can accentuate this problem. Such contracts are designed for situations characterised by high professional standards and competence and adequate related capacity. In contexts where such standards and competence are not evident, and/or where there are related capacity constraints, they can give rise to corruption risks. These may have been avoidable under alternative forms of contract based around a more collaborative approach to managing risk and reward.

2.3.5 Relevant International Standards

2.3.5.1 ISO 10845

ISO 10845 is the international standard for construction procurement. Published in 2010, it describes processes, methods and procedures for the establishment within an organisation of a procurement system that is fair, equitable, transparent, competitive and cost-effective (ISO, 2010). This detailed set of documents is based around facilitating two sets of objectives relating, respectively, to: good governance and the use of procurement to promote social and national agendas. This is potentially well suited to the transparent embedding of a socially and environmentally sustainable development agenda within the process of infrastructure procurement.

From an anti-corruption perspective, a notable feature of ISO 10845 is its practical focus on building trust between all those with an interest in procurement (Watermeyer, 2012). In addition, it sets how to deal with challenging situations in a rational and transparent manner that provides reassurance to all involved, thus reducing associated corruption risks.

Even when other procurement systems have been adopted, ISO 10845 can serve as a helpful point of reference, or checklist, when exploring ways of addressing apparent shortcomings in those other systems.

2.3.5.2 ISO 37001

ISO 37001 is the international standard for anti-bribery management systems (ISO, 2016). Its role in managing corruption risks derive from both its approach and its scope. Being systems-based, it requires strong internal risk assessment and performance monitoring systems that measure success and failure, so that lessons are learned and policies, procedures, training and practices are continually reviewed and periodically updated.

ISO 37001 does not itself provide policies and procedures to be adopted by organisations. Rather it sets out requirements against which an organisation's own policies, procedures and practices can be assessed. As such, its adoption requires a high level of management engagement, thus reducing the risk of token compliance.

2.3.5.3 OC4IDS

The Open Contracting Data Standard (OCDS) is a globally recognised open data standard for public procurement. Already adopted by over 30 governments around the world, it describes from a technical perspective how to publish data and documents related to the procurement of goods, works and services. When it comes to the particularly challenging matter of infrastructure procurement, the Open Contracting for Infrastructure Data Standard (OC4IDS) is recommended. This describes how to consistently combine contract-level disclosures (using the OCDS) and project-level disclosures (based on the CoST Infrastructure Data Standard) to support scalable disclosure and monitoring of infrastructure projects (Open Contracting Partnership, 2019).

2.3.5.4 Quality infrastructure investment (QII) principles

Published in 2019, the G20 QII principles comprise six voluntary, nonbinding principles that provide a strategic direction for infrastructure investment (World Bank, 2023). These build on the consensus that infrastructure is a significant driver of economic prosperity, and that well-built and sustainable infrastructure maximises the positive impacts of these high-priced investments. They can be summarised as follows:

- Maximising the positive impact of infrastructure to achieve sustainable growth and development
- Raising economic efficiency in view of life-cycle cost
- Integrating environmental considerations in infrastructure
- Building resilience against natural disasters
- Integrating social considerations in infrastructure investment
- Strengthening infrastructure governance

Together with associated guidance and indicators, these are clearly relevant to the formulation and implementation of approaches to the planning, provision and use of infrastructure that contributes to sustainable development. Principle 5 includes a particular focus on the importance of inclusive participatory approaches, while Principle 6 highlights the need for information disclosure, transparency and accountability measures.

2.3.6 The Commonwealth Anti-corruption Benchmarks

Published in 2021 and informally endorsed by the leaders of all 54 Commonwealth countries, this comprehensive document provides 25 benchmarks of good anti-corruption practice for national governments and public sector bodies (ComSec, 2021b) – see

FIGURE 2.2 Commonwealth anti-corruption benchmarks.

Figure 2.2. Designed to be relevant to *all* countries, the benchmarks are first summarised, and then specified, with detailed associated guidance provided in supporting Annexes.

Together with supplementary documents including a road map for adoption, and checklists for monitoring progress, this constitutes a valuable toolkit, not only for governments seeking to curb corruption risks, but also for those seeking to hold governments to account.

The benchmarks are aspirational. No country yet meets all of the specified requirements, though some have chosen to embark on the process of doing so.

The Global Infrastructure Anti-Corruption Centre (GIACC, 2023) was responsible for the process of preparing these resources in close consultation with the African Union, the IMF, United Nations Office on Drugs and Crime, Commonwealth law ministries, anti-corruption agencies and partner organisations. Though all the benchmarks are important, several are particularly relevant to curbing corruption risks related to public infrastructure. These include those related to public sector organisations (BM10), public officials (BM11), procurement (BM13), contract management (BM14), anti-corruption training (BM20), reporting corruption (BM21), professional institutions and business associations (BM23) and participation of society (BM24).

2.3.7 Lessons from Health and Safety

In the 1990s, the construction sector in many countries had a reputation not only for corruption, but also for poor Health and Safety standards. Though construction sector fatalities and injuries remain high compared with other sectors, they have since reduced markedly. This can be attributed to a combination of severe penalties at board level for non-compliance with regulations, and the adoption of a systems approach whereby lessons are routinely learned from near misses as well as actual instances of injury.

The construction sector has also seen the introduction of severe penalties at board level for proven cases of corruption. But more often than not policies of "zero-tolerance" of corruption appear not to be supported by a systems approach that explicitly seeks to learn lessons from near misses. As a result, staff who are aware of such risks may choose to turn a blind eye to them rather than risk disruption to their work or that of others by raising topics about which they know there to be "zero-tolerance" (Goldie-Scot, 2023).

2.4 CORRUPTION MANAGEMENT

2.4.1 Contrasting Approaches to Corruption Risk Management

Reviews of successful anti-corruption reforms have identified little if any evidence that legal and institutional reforms in themselves have any attributable positive impact (DFID, 2015). Examples abound where applicable laws are in place, but are either not enforced at all, or enforced in a manner that is itself characterised by corruption. By contrast, there are cases where countries or sectors enjoy low perceived levels of corruption, despite appropriate laws not being in place, and/or there being very little capacity to detect wrongdoing and bring prosecutions. The UK is a case in point. During the period before the Bribery Act, perceived levels of corruption were low despite (according to the OECD) a lack of appropriate legal provisions.

An important consideration when formulating an anti-corruption strategy is therefore the determination of the appropriate balance between (broadly speaking "top-down") approaches primarily based on laws and government institutions, and systems-based approaches that are broadly speaking "bottom-up" and emphasise the importance of collaboration between stakeholders.

Characteristic features of the two approaches are summarised in Table 2.1. Neither approach is inherently "right" or "wrong" as such. Rather, a balance of the two is required. That said, corruption risks associated with an unduly top-down approach tend to be particularly high on account of the incentives that can arise for all involved to maintain a veneer of compliance, while turning a blind eye to malpractice or inefficiency when it occurs. This is particularly the case in authoritarian contexts where even the "bottom-up" features such as transparency and participation are unduly influenced by deference to authority.

Singapore is unusual in this regard in that it achieved, and has sustained, its status as the least corrupt country in Asia through the adoption of what would in most other contexts considered to be an inefficient and therefore ultimately unsustainable top-down approach. Its success has been attributed to the genuine political will of the government and the allocation of adequate resources to the Corrupt Practices Investigation Bureau to ensure that it enforces anti-corruption laws impartially, regardless of the offender's position, status or political affiliation. Key lessons from Singapore's success (Quah, 2016) in addressing corruption risks include the importance of the following:

TABLE 2.1 Characteristics of contrasting anti-corruption strategies

	LEGALISTIC APPROACH	*COLLABORATIVE SYSTEMS APPROACH*
Summary	Overall objective is "anti-corruption"	Overall objective is "good performance"
	Regulations are inherently good	Regulations are inevitably insufficient
	Deviations from official processes and procedures are not tolerated	Processes and procedures can always be improved
	Discourages honest assessment and discussion of risks	Encourages honest assessment and discussion of risks
Typical features	Generally led by lawyers	Generally led by managers or engineers
	Imposed	Collaborative
	Identifies and punishes non-compliance	Seeks to understand and address issues
	All is well unless rules are broken	Risks are widespread and unpredictable
	Detailed policies and procedures intended to cover any possible scenario	General principles and clear processes
	Discourages discretion	Allows discretion, with accountability
	Suppresses open discussion of real issues	Encourages transparency and frank discussion
	Adds further layers of control each time something goes wrong	Improves processes in response to near misses as well as incidents
	Direct focus on legal compliance	Direct focus on professional integrity
	Trains staff to learn the rules	Trains staff to understand and address risks

1. Genuine political will.
2. Institutional clarity and capacity through a single well-resourced agency.
3. Impartiality, in that the agency has not been used as a weapon against political opponents.
4. Focus on underlying drivers. The agency seeks to systematically understand and address the causes of corruption, rather than relying on what it views as ineffective and discredited "anti-corruption" campaigns.

2.4.2 Understanding and Addressing the Causes of Corruption

To be effective, any attempt to curb corruption, or indeed improve performance, must meet two basic requirements. Firstly, the approach must be informed by a model of the system that it is seeking to regulate, with the model possessing the same core degrees of freedom as the system it is seeking to model (Ashby, 1970). Secondly, those seeking to achieve change must communicate a vision for change through a concept or formula that is capable of being widely understood and accepted.

As already described, the most commonly used model can be summarised in the notion of "zero tolerance". This views corruption as being a deviation from acceptable practice, and typically the exception rather than the rule. According to this model, an appropriate response to corruption is to determine where it has occurred, and then either to strengthen the institutional and legal environment to ensure that such behaviour is criminalised or – if the necessary laws and regulations are already in place – to enforce them more vigorously. Such an approach may be helpful or even promising at times but is certainly not strategic if it does not necessarily address underlying drivers of behaviour.

In terms of an overarching formula or concept to communicate the approach, Klitgaard's C=M+D−A formula (Corruption=Monopoly+Discretion−Accountability) (Myers Fletcher & Gordon, 2022) was initially accepted by MDBs, but in practice soon become simplified to a combination of C=M and C=D. In other words, the way to minimise corruption risks came to be seen as being a combination of minimising any scope for discretion (decisions by individuals to deviate from set procedures) or monopoly (restrictions on competition).

By diluting Klitgaard's acceptance of discretion with associated accountability, this approach has in many cases given rise to a culture of token "compliance" rather than considered, and accountable, optimisation of approach. Particularly in the case of large institutions, this may be understandable in that clear rules need to be established. However, it is ultimately only fully effective in practice if policies and procedures are so well formulated that they are capable of taking into account every conceivable scenario that officials, or staff, may face. Realistically this is rarely if ever the case.

An innovative and seemingly more effective way of mapping, understanding and addressing the drivers of corruption risks is to focus instead on any apparent shortcomings in the drivers of performance. First tested in 2008 in the course of a study of construction sector corruption risks in Ethiopia (World Bank, 2012), the "ACTS" (Accountability Capacity Trust Setting) approach has since been successfully applied in many countries, and has been adopted by the Infrastructure Transparency Initiative (CoST) as a tool to help independent assurance teams identify and better understand likely areas of corruption risk (CoST, 2020a).

For any sector, at any level, each aspect of performance in the planning and provision of goods, services and works must be conducted in a manner that demonstrates quality in technical, social, economic and environmental terms. The ACTS approach and associated Excel tools (CoST, 2020b) consider this to require a combination of the following drivers:

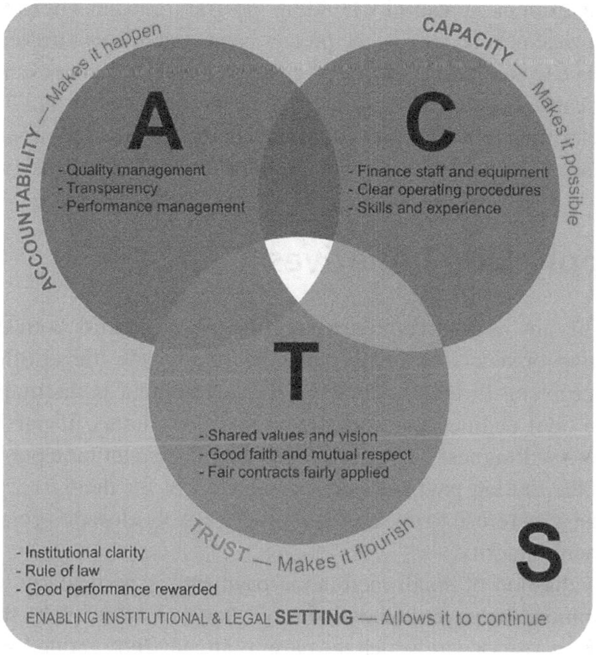

FIGURE 2.3 Drivers of good performance (CoST Assurance Manual, 2020a).

1. The **Accountability** mechanisms to make it *happen*
2. The prerequisite **Capacity** to make it *possible*
3. **Trust** in the fairness of the procurement system to allow it to *flourish*
4. Enabling institutional and legal **Setting** to allow good performance to *continue* in the long term

As illustrated in Figure 2.3, each of these drivers can in turn be broken down further into its constituent elements. Where any one of those elements is absent or weak, that is considered to constitute a management failing that simultaneously gives rise to a risk of corruption. Looked at this way, this tool serves to help

1. map potential corruption risks,
2. illustrate the need for a broad-based approach to improving sector performance,
3. highlight the importance of trust, and
4. identify constructive ways of bringing stakeholders together to work collaboratively in improving sector performance.

When all the drivers are in place, the "sweet spot" in the centre of the figure is achieved. This can result in a marked improvement in performance, whether at the level of a contract, a project or a whole sector.

This approach also highlights the futility of isolated ad hoc "anti-corruption" interventions that either fail to take account of, or unwittingly undermine, other drivers of good performance. For example, a "capacity-building" programme cannot be expected to enhance private sector engagement so long as the procurement system is not trusted, and/or it is not possible in practice to enforce a contract. By contrast, it soon becomes clear that an effective way of building capacity within a sector may not be through capacity-building interventions *per se,* but by reforming procurement systems and the institutional environment to the point that the private sector has sufficient confidence to invest in its own capacity.

2.4.3 Supply-Side Initiatives

To be successful, any anti-corruption strategy needs to address both the demand-side (those demanding or receiving bribes) and the supply-side (those offering or paying bribes) of the problem. In many countries, the government is the main, or even only, client for some civil engineering contractors and consultants. If participation in that market requires a willingness to pay bribes (as an inducement for a public official to act improperly) or facilitation payments (as an inducement for them to act properly when they would otherwise refuse to do so), then it is effectively closed to companies committed to professional integrity.

The criminalisation of small facilitation payments is particularly unhelpful in this regard. By definition, the payer is the victim of a form of extortion by the public official who is withholding a service for which the payer is already fully entitled. A fairer approach would arguably be to extend the permissible mitigating circumstances to include companies that resist attempts to extort facilitation payments but, if necessary, go ahead and make the payment, while doing so in a manner that is transparent and fully documented. By contrast, the official demanding the facilitation payment would still be committing an offence.

International Federation of Consulting Engineers (FIDIC) has extensive experience in developing tools and resources to help its members act with professional integrity. In early 2001, it published the Business Integrity Management System as a standard for adoption by private companies. However, its take-up was curtailed after both the World Bank and the European Commission ruled that it would be anti-competitive for clients to use its adoption as one of the bid evaluation criteria.

In 2011, FIDIC launched the FIDIC Integrity Management System (FIMS). This was again focussed primarily on the supply-side of corruption risk management, so had only limited impact, given that many of the companies that are part of the FIDIC member association work in markets in which it is not possible to participate with complete integrity. In 2019, further guidelines were published (FIDIC, 2019a) to help companies that already have a functioning FIMS move towards the adoption and eventual certification of an integrity management system incorporating an anti-bribery framework under the ISO 37001 Standard.

2.4.4 Lessons Learned

A great deal has been learned in the 25 years or so since it was generally accepted that corruption has a negative impact on development. As already outlined in this chapter:

1. Institutional, legal and regulatory change is important but not sufficient. In many cases it does not help at all, and in some cases can make things worse.
2. Many if not most people engaged in or turning a blind eye to corruption consider that they have no choice.
3. If there is to be any realistic prospect of change, it is necessary to address the demand-side and the supply-side of corruption risks in parallel in a coherent manner that is informed by a clear understanding of underlying drivers of behaviour.
4. A systems approach is required, so that lessons can be learned from near misses, and appropriate action taken to understand and address weaknesses in current processes. For this to be realistically achieved, there should be a recognition that corruption risks abound, and that there is always scope for improvement. This means that staff should be encouraged to talk about corruption risks without fear of being criticised for doing so.

Distilling this further, it is helpful in any given situation to assume as a starting point that everyone is to some degree compromised by corruption risks, but equally that everyone is also a potential ally in better understanding, and thereby addressing, such risks.

2.4.5 The Infrastructure Transparency Initiative (CoST)

The Infrastructure Transparency Initiative (CoST, 2023a) adopts just such a collaborative approach. CoST members help stakeholders from government, the private sector and civil society work together to let the facts speak for themselves. This is most notably achieved through the CoST "assurance" process whereby data are disclosed in accordance with an agreed standard and turned into compelling information that is relevant to the needs of all stakeholders. Crucially, the resulting reports can be trusted, because they are not generated by government, or the private sector, or by civil society, but by a formally established multi-stakeholder group that acts as the board responsible for all activities under a CoST programme, which may operate at a national, subnational or mega-project level.

The four features of CoST are illustrated in Figure 2.4. An important aspect of the approach is that its aim is to strengthen existing accountability mechanisms, and not to replace or otherwise undermine them. The CoST approach can therefore be summarised as follows:

- *working together*; in order to,
- *let the facts speak for themselves*; in order to,
- *strengthen existing accountability mechanisms*; in order to,
- *improve performance.*

CoST assurance entails a team of independent experts, including civil engineers, being appointed by a multi-stakeholder group to examine all available official data and generate findings and recommendations. The ACTS tool is used to help map potential corruption risks and thus inform subsequent data analysis and stakeholder consultation.

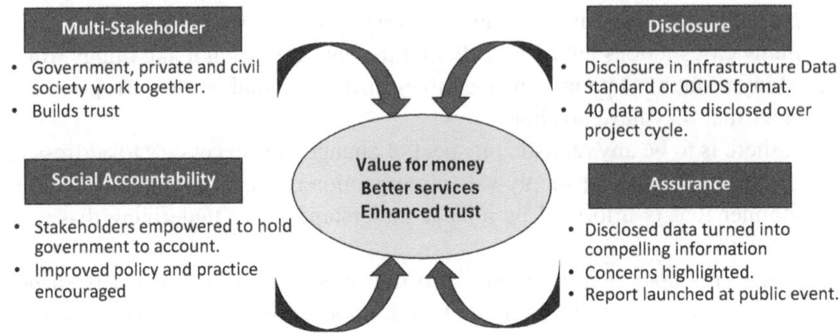

FIGURE 2.4 The four features of the CoST approach.

Without pointing a finger of blame, the resulting recommendations not only address issues on specific projects but also facilitate broader improvements in sector governance. These reports can then be used by anyone, whether government officials, media organisations, civil society groups or others to ask pertinent questions and help hold decision-makers to account.

There is growing evidence (CoST, 2023b) of remarkable success of this approach in several of the countries in which the CoST approach is currently applied. By catalysing improvements in competence, capacity and trust, CoST helps generate savings in both time and cost. In Thailand, the total value of projects subjected to the CoST approach had by 2022 reached over US$ 6 billion. The availability of data from a set of equivalent "non-CoST" projects led to an independent estimate of cost savings of about 12% attributable to the deterrent effect of the increased scrutiny (where actual or potential) associated with the CoST approach.

Features of the CoST approach that differentiate it from other similar initiatives are as follows:

- It simultaneously addresses supply-side and demand-side corruption risks
- Its constructive nature, explicitly aimed at building trust between stakeholders
- Its project-level focus, drawing from different data silos
- Its principle of not duplicating or undermining existing accountability mechanisms.

FIDIC has endorsed CoST (FIDIC, 2019b) for being consistent with its own commitment to quality, integrity and sustainability. CoST also appears in the G20 Compendium of Good Practices for Promoting Integrity and Transparency in Infrastructure Development (G20, 2019).

2.4.6 The Global Procurement Partnership (GPP)

Elements of the foregoing are to some extent reflected in the work of the Global Procurement Partnership Initiative. Led by the World Bank and launched in 2016 with strong Swiss support, this aims to advance worldwide, and country-level procurement reforms increase coordination and maximise impact. In early 2022, the GPP published a major report (World Bank, 2022) focussed on sustainable development.

Positive features of this report include its recognition of the risks of neglecting the planning and contract implementation stage of procurement, and of the potential for public procurement to be used as a vehicle for promoting aspects of public policy related to sustainability and social inclusion.

This report acknowledges the limited success of past procurement practices in achieving value for money. It does not, however, seek to identify, understand and address possible underlying causes. Instead, it is focussed primarily on introducing fresh policy imperatives, most notably those related to sustainability and the protection of disadvantaged groups through the introduction of privilege-resistant policies.

The GPP nevertheless remains well placed to potentially serve as a vehicle for widespread procurement reform that is informed by an understanding of underlying drivers of performance as portrayed in the ACTS framework, as well as by the positive impact of the CoST approach in a wide range of contexts.

2.5 CONCLUSION

Despite significant progress in regulatory reform and related compliance measures over the past 25 years, corruption risks remain high in relation to the planning and provision of public infrastructure. Not only does this aggravate the infrastructure funding gap, but related shortcomings in governance and quality management also undermine efforts to strengthen associated sustainability considerations.

Part of the reason for this lack of progress is that attention continues to be unduly focussed on symptoms rather than on underlying causes. Unless very carefully formulated, approaches based on "zero tolerance" of deviation from prescribed norms can be unduly one-dimensional. These can then become characterised by a lack of flexibility, multiple layers of accountability, and over-reliance on people both understanding, and being realistically capable in practice of adhering to the associated plethora of rules and regulations that cannot, in practice, realistically cater for all possible scenarios.

Though there is no single, or simple solution, there is growing evidence that corruption risks can be better understood and mitigated through approaches that are instead characterised by (1) government working in genuine partnership with civil society and the private sector to understand the underlying drivers of corruption risks and ensuring that regulations intended to deter corruption are likely in practice to do so; (2) transparency that gives rise to related accountability by turning objective data into information that is relevant to the needs of, and trusted by, each group of stakeholders; (3) effective quality assurance throughout both the project cycle and the associated procurement cycles, with the concept of "quality" sufficiently broad to take account of social and environmental factors as well as technical and economic ones; and (4) a systems approach that encourages the ready acknowledgement of mistakes, so that lessons are learned and appropriate corrective action taken.

Most of these characteristics are already to some extent reflected in different aspects of recognised good practice. However, they are not generally brought to bear in a coherent manner. This is partly because of a perception that progress is being made simply by keeping busy dealing with, or studying, the symptoms of corruption, and partly because of the absence of an overarching strategic narrative to which decision-makers can refer.

A practical example of such an overarching narrative is the ACTS framework. As detailed in this chapter, this considers the overlapping drivers of good performance to be the Accountability mechanisms needed to make good performance happen, the Capacity needed to make it possible, the Trust needed for it to flourish, and the institutional and regulatory Setting needed for it to continue on a sustainable basis.

If such lessons, and others, can be learned, refined and communicated, the prospects for early demonstrable impact appear good. This is borne out by the experience of the CoST Infrastructure Transparency Initiative. This, though necessarily still tentative and relatively small scale, has started to generate an evidence base of the latent potential of approaches based on a combination of constructive engagement, transparency and strategic evidence-based policy development.

REFERENCES

ADB (2005) *Special Evaluation Study: The Role of Project Implementation Units*. The Asian Development Bank. Available at: https://www.adb.org/documents/project-implementation-units

Ashby, R. C. (1970) Every good regulator of a system must be a model of that system. *International Journal of Systems Science*, 1(2), 89–97.

Business Insider (2020) *Berlin's New $7 Billion Airport has Finally Opened after 9 Years of Delays, Corruption Allegations, and Construction Woes*. Available at: https://www.businessinsider.com/berlin-brandenburg-airport-finally-opens-after-9-years-of-delays-2020-11?r=US&IR=T

Cole, W. M. (2015) Institutionalizing a global anti-corruption regime: Perverse effects on country outcomes, 1984–2012. *International Journal of Comparative Sociology*, 56(1). 53–80.

ComSec (2021a) *Commonwealth Advance Online Training in Government Performance Management*. Available at: https://production-new-commonwealth-files.s3.eu-west-2.amazonaws.com/migrated/inline/Concept-Note%20%28Africa%20Europe%20Edition%29%20Jan%202021%20%28FINAL%29.pdf

ComSec (2021b) *Commonwealth Anti-Corruption Benchmarks*. Available at: https://production-new-commonwealth-files.s3.eu-west-2.amazonaws.com/s3fs-public/2022-04/Anti-Corruption%20Benchmarks_UPDF.pdf?VersionId=mDx64yykQWIJ_T.XhdkOnEtyf_NH_n4b

CoST (2020a) *Assurance Manual*. Available at: https://infrastructuretransparency.org/resource/assurance-manual/

CoST (2020b) *Assurance Manual Annexes: Excel Tools*. Available at: https://infrastructuretransparency.org/resource/assurance-manual-annexes-excel-tools/

CoST (2023a) *CoST Impact Stories*. Available at: https://infrastructuretransparency.org/costimpact/

CoST (2023b) *Our Approach*. Available at: https://infrastructuretransparency.org/our-approach/

DFID (2015) *Why Corruption Matters: Understanding Causes, Effects and How to Address Them.* Available at: https://assets.publishing.service.gov.uk/government/uploads/system/uploads/attachment_data/file/406346/corruption-evidence-paper-why-corruption-matters.pdf

FIDIC (2019a) *Guidelines for Integrity Management in the Consulting Industry.* Available at: https://fidic.org/sites/default/files/FIMS%20III_OCTOBER%202019_FINAL%20VERSION.pdf

FIDIC (2019b) *FIDIC Endorses Global Infrastructure Transparency Initiative Cost.* Available at: https://fidic.org/node/26814

Friedmann, G. (2009) *The Next 100 Years.* Doubleday, New York.

G20 (2019) *Compendium of Good Practices in Infrastructure Development.* Available at: https://www.oecd.org/g20/summits/osaka/G20-Compendium-of-Good-Practices-in-Infrastructure-Development.pdf

GIACC (2023) *Global Infrastructure Anti-Corruption Centre.* Available at: https://giaccentre.org

Goldie-Scot, H. (2020). *Why Infrastructure Transparency Matters.* Available at: https://www.ice.org.uk/news-insight/news-and-blogs/ice-blogs/the-civil-engineer-blog/why-infrastructure-transparency-mattersGoldie-Scot, H. (2023) *Surveys during Anti-Corruption Training Courses for Staff Working in Organisations with a "Zero Tolerance" Policy towards Corruption.*

ICE (2022) *Code of Professional Conduct.* Available at: https://www.ice.org.uk/download-centre/code-of-conduct

IMF (2020) *Well Spent: How Strong Infrastructure Governance Can End Waste in Public Investment.* International Monetary Fund, Washington, DC.

ISO (2010) *ISO 10845-1:2010 Construction Procurement.* Available at: https://www.iso.org/standard/46190.html

ISO (2016) *ISO 37001:2016 Anti-Bribery Management Systems.* Available at: https://www.iso.org/standard/65034.html

Klitgaard, R. (1991) *Controlling Corruption.* University of California Press, Oakland.

Lomborg, B. (2022) *Good News, the World is getting better.* American Institute for Economic Research, Great Barrington, MA.

Marshall, T. (2023) *The Future of Geography.* Elliott & Thompson, London.

Marx, K. and Engels, F. (1848) *Manifesto of the Communist Party.* Available at: https://archive.cpgb-ml.org/download/publications/marx_engels_communistmanifesto.pdf

Myers Fletcher & Gordon (2022) *The Formula for Corruption in Construction.* Available at: https://myersfletcher.com/the-formula-for-corruption-in-construction/

Napier (2022) *11 of the Biggest FinCrime and Money Laundering Facts.* Available at: https://www.napier.ai/post/financial-crime-statistics-2022

Neary, A. (2026). *Rock Springs landfill,; first in world diversion of E-waste using pyrolysis process.*" Sweetwater County, Wyoming, USA.

OECD (2008) *OECD Group Demands Rapid UK Action to Enact Adequate Anti-Bribery Laws.* Available at: https://www.oecd.org/daf/anti-bribery/oecdgroupdemandsrapidukactiontoenactadequateanti-briberylaws.htm

Open Contracting Partnership (2019) *Open Contracting for Infrastructure Data Standards Toolkit.* Available at: https://standard.open-contracting.org/infrastructure/latest/en/

Pw C. (2014) *Economic Crime: A Threat to Business Globally.* Available at: https://www.pwc.com/gx/en/economic-crime-survey/assets/economic-crime-survey-2014-construction.pdf

Quah, J. S. (2016) Singapore's success in combating corruption: Four lessons for China. *American Journal of Chinese Studies*, 23(2):187–209.

Reuters (2019) *Canada's SNC-Lavalin Unit Pleads Guilty to Fraud Charge in Libya Case.* Available at: https://www.reuters.com/article/us-snc-lavalin-court-idUKKBN1YM1V0

Runciman, D. (2018) *How Democracy Ends.* Profile Books Ltd., London.

Transparency International (2023) *What is Corruption?* Available at: https://www.transparency.org/en/what-is-corruption

Tucker, S. C. (1999) *Vietnam*. Warfare and History Series. UCL Press, London.

UN (2025) *UN Population Division Data Portal*. Available at: https://population.un.org/dataportal/home?df=a5a8b6db-b9dc-46c3-a7df-9892d53259c3

UNODC (2023) *Money Laundering*. Available at: https://www.unodc.org/unodc/en/money-laundering/overview.html

World Bank (2005) *Voice for the World's Poor: Selected Speeches of World Bank President James D Wolfensohn*. Washington DC. Available at: https://www.worldbank.org/en/archive/history/past-presidents/james-david-wolfensohn

Watermeyer RB (2012) A framework for developing construction procurement strategy. *Proceedings of the Institution of Civil Engineers – Management, Procurement and Law*, 165(4), 223–237.

World Bank (2012) *Diagnosing Corruption in Ethiopia: Perceptions, Realities and the Way Forward for Key Sectors*. Available at: https://documents1.worldbank.org/curated/en/210171468024611636/pdf/699430PUB0Publ067869B09780821395318.pdf

World Bank (2022) *A Global Procurement Partnership for Sustainable Development*. Available at: https://documents.worldbank.org/en/publication/documents-reports/documentdetail/173331642410951798/synthesis-report

World Bank (2023) *The QII Principles in Action*. Available at: https://www.worldbank.org/en/programs/quality-infrastructure-investment-partnership/qii-principles

Younger, J. S., Booth, D. J., and Kurniawan, K. (2012) Sustainable development: East Bali Poverty Project. *The Proceedings of the Institution of Civil Engineers-Municipal Engineer*, 165(1):43–58.

Education and Poverty

3

Scott Younger, Azra Meadows, and David Booth

3.1 EDUCATION ISSUES

3.1.1 Education Overview

Brown (2021) has drawn attention to the lack of education among the world population with three-quarters receiving a substandard education and, regrettably, too many impoverished with no access to education at all. In round numbers, this means that about 1.6 billion are illiterate, more or less, 4.4 billion in the world fall into the category of inadequately educated, and only 2 billion have an adequate to fulsome one. He demonstrates clearly how education is inextricably linked to poverty and the opportunities to climb out of it. Accordingly, this chapter presents details of projects which show how, with the correct dedication to education, poverty can be overcome permanently.

About 60 years ago, there were only 3 billion people in the world, and the numbers in poverty were far lower, which demonstrates the stark reality of the current situation and the huge and growing disparity between those who have benefitted from the opportunity for a fulsome education and those for whom the chance remains a dream. This is complicated by the fact that there are significant differences between countries. These differences depend on physical size, climate, history, population and their level of development or how advanced they have become in terms of governance, societal factors, politics and corruption and how it is dealt with.

Meeting these education challenges is vital for the future of the sustainability of the human race. For this, there are two extremes which we must note, and there is a large difference between them which is necessary for the education of the vast majority of people. At the bottom extreme, we have to deal with the education of people living in

DOI: 10.1201/9781003566274-3

poverty, some illiterate, while at the top end, we will have people at the cutting edge of knowledge, looking for tertiary-level training and more learning. This will include the various ramifications of artificial intelligence (AI), for which scientists and the curious population are already showing considerable interest. Space exploration also, after an absence of some three decades, contributes significant educational spin-offs. The large majority of people can be handled educationally in a standard manner, except that the teaching today is largely inadequate for the vastly expanded numbers, and generally poorly delivered. There needs to be an upgrade of standards for which there is a wide range in delivery across the countries, developed ones included, not to mention where other factors, such as out-of-date religious beliefs, are obstructing the teaching of girls.

In this discussion of education and poverty, rather than a generic portrayal, case studies are taken to cover a range of problems. In the first and completely comprehensive situation, a totally 'lost' community of 17,000 people living in a mountainous environment is taken as an example. At this very bottom level of society, the example presents a project, which highlights many of the large range of problems faced when dealing with lifting a 'forgotten' district out of poverty and towards a good level of sustainability, thereby giving the communities the confidence and positivity to contribute to society.

3.1.2 Education of the Highly Skilled

Macaskill (2023) has written about his concern for having skilled people around to ensure the research and developments for the long-term future of the world. He discusses eloquently the issues that are faced and fears catastrophes and long periods of stagnation with parallel decline in numbers of people, leading to the collapse of civilisation. The tenor of his book is that we must get through this century to ensure sustainability, which is deemed to be difficult enough, as we battle through many problems. There are the issues, raised by Brown (2021), regarding the unsettled political situations in several regions, which show no sign of reducing. In addition we need to deal with the aftermath of a pandemic associated with a modifying and not yet defeated virus.

The political situation is complex, perhaps more so because of the greater number of people who have to have their needs addressed. World leadership tends to go in cycles, and countries have to react to faster-moving changes. It would appear that the world has entered another period of change. Runciman (2018) believes the advent of Trump as president of the USA, and this was just an observation in his first term, signalled the end of democracy. If so, what will replace it, and what will happen with education?

3.1.3 Advances in Education

The population at mid-century is forecast to be approximately 11.0 billion (see Figure 1.2). This means nearly an additional 3 billion people to be catered for, compared with today's population. However, the great majority of this additional population will be in Africa. Therefore, there would appear to be a huge world need to turn urgent attention to the countries in Africa. This is likely to happen and has already started. In addition to having the highest population growth, many African countries are among the poorest in the world.

Language is an important issue in education and knowledge transfer. The UK, North America and Australasia and several other countries benefit from English being their first language. It is also the acknowledged language to be acquired as a second tongue, if it is not the first. India has a large percentage of English language speakers because of its past links to the UK, and several countries are also linked through the British Commonwealth. Other widely spoken languages are Spanish, as used mainly in Central and South America, and Chinese in Asia, spoken in several dialects.

A new look at education learning was promoted in Boston, USA, about a decade or so ago. It felt that education was not adequately developing the ingenuity or creativity of students in the classroom. They called the new approach Nu Vu education, and secondary-level students are taught to think about the real world in a studio environment. The first school in the UK to embrace this approach in 2019 was the 145-year-old Kelvinside Academy in Glasgow. The installation of the facility has raised interest in the industry in the city, and they are encouraged to use the facility in times when it is not in use by the school.

Of course, the subject which is catching the most active interest today is AI. The use of AI can be open to abuse, however, and while the subject is also given to widespread disinformation, there is no doubt that AI is here to stay, is developing quickly and can be a very useful tool when applied properly (Kirwan, 2024). In a few years, it can be expected that all school leavers will at least have a basic knowledge of its capabilities.

Turning attention to training in its broadest sense. Education was the theme of some of the aid-funded projects undertaken in Pakistan, led by Peter and Dr Azra Meadows of Glasgow University. The detail is included alongside a number of aid-funded projects, which focused on the environment and poverty, and various projects were undertaken across the nation. This is the most conventional way in which to provide assistance, that is, to provide appropriate expertise to an existing arrangement to either introduce new ideas or focus in order to enhance the abilities of in-place staff. One project involved teaching of 'green' skills for handicrafts to villages which had a high proportion of beggars, people with little or no income. This was highly successful, and the improvement in the self-dignity of the beggars was pleasing to see. They were at last contributing to their community albeit at a lower level.

3.2 EDUCATIONAL EXAMPLES

3.2.1 INSPIRE – Transnational Education Programme

Indus River natural resources management: The Higher Education Commission and British Council Pakistan funded a 1-year (2012–2013) career development programme, which was executed between Shah Latif University, Sindh and Glasgow University. This led to a 3-week Career Development Course focused on the Indus River and the management of its resources and ecosystem services and conservation strategies. It included consideration of the range of Indus River ecosystems that rural communities directly or indirectly depend on as a means for food, shelter and livelihoods. It also included environmental impact assessments (EIAs) of the risks and mitigation of natural disasters with such as periodic flooding and earthquakes, biodiversity conservation and the design of eco-friendly hazard-resistant buildings.

The course on 'Indus River Natural Resources SGIDF Management' was highly relevant to the development priorities of the country. These priorities are included in Pakistan's Interim Poverty Reduction Strategy Paper (I-PRSP 2001, 2003), the Asian Development Bank's (ADB's) Country Strategy Plan for Pakistan (2005) and the priorities of the Higher Education Commission of Pakistan. They also cover the country's research priorities, teaching development and quality assurance. The 3-week course covered a range of generic skills. These included data collection and analysis, parametric and non-parametric statistics, field survey methodologies, report preparation and presentation.

Participants were expected to have basic computer skills of word processing and spreadsheets and were shown to use a statistical package during the second week of the course. Field work consisted of surveys of relevant ecosystems that were accessible from Shah Abdul Latif University. The sites chosen were from those at which the Heritage Foundation and the Pakistan Wetlands Project have ongoing programmes. These involved ecological surveys, wetland management, biodiversity conservation, ecosystem services, rural community uplift and eco-friendly flood-resistant housing. Field work also assessed permaculture methods used by rural communities to provide small-scale agricultural systems that are sustainable in the long term.

3.2.2 Developing Inclusive and Creative Economics

Green skills and crafts for livelihoods: This 1-year (2019–2020) British Council-funded project built upon the experience and knowledge gained over several years in working with marginalised sections and climate migrants by both the Heritage Foundation and Glasgow University. The purpose of this project was to support barefoot social and creative enterprises in eight villages in the vicinity of the World Heritage site at Makli, Sindh Province.

The enterprises consisted of eight categories and 25 subcategories of activities, designed to provide livelihoods to marginalised sections of the communities, mainly consisting of beggars attached to the villages. The project has helped several village communities to give up begging and start to live a life of some dignity. The creation of the proposed structure helped to build centres of specialist products (one village/one product). It provided an opportunity for each village to specialise in selected categories of green/sustainable/organic activities along with cultural, creative output based on improved craft-making developed from conservation work carried out at the World Heritage site. Since the outputs from the activities were local, sustainable and low cost, these were within the affordability of impoverished households. This project was considered as a successful pilot in the quest for providing better services and activities for the benefit of low-income populations living at the lowest levels of society. In view of its long association with the Heritage Foundation, the review and monitoring by the Glasgow University staff were invaluable in creating an improved structure, which could be replicated in Pakistan and elsewhere.

3.2.3 Tertiary Educational Example

A further example at the tertiary level was to introduce a whole university programme, in this case a UK Master's level programme in Highways and Transportation, into the Institute of Technology Bandung Indonesian (ITB), one of the country's top universities. This was introduced in the mid-1980s in Indonesia with specialist staff engaged, while ITB counterpart staff members received appropriate training and further degrees in the UK in order to be able in time to take over. This required about 10 years of a few full-time plus a number of part-time specialist overseas staff members, in addition to UK university affiliation, Leeds and Strathclyde in this case, to provide accreditation with the Ministry of Education (Younger, 1992).

3.2.4 Lessons to Be Learnt

The above sections demonstrate that the provision of education, not to forget training, is wide ranging, required at many levels, from preschool training all the way to tertiary university learning. It is required at an expanding rate, in particular, at the early stages, and there is an opportunity to do it correctly in the areas of burgeoning population growth, such as being experienced in Sub-Saharan Africa. It is vital that local staff are included in the teaching, as required, as they will be expected to take over and run the programmes themselves.

The overriding lesson from the experiences of the past is that education is most successfully carried out when a community is involved. It shows the value of smaller units of collegiate responsibility.

3.3 DEALING WITH POVERTY

3.3.1 General

This section concerns the critical issue of poverty and its relief, in which education can play a prominent part and without which the other elements to alleviate poverty would likely not be effective, and the community being helped would gradually go into further decline.

Poverty has been around for a very long time, and a solution to this fundamental problem of human existence would appear to be as far away as ever. It is naturally at the top of the agenda of the current UN's Sustainable Development Goals (2015–2030) – as shown in Table 1.1 – and, with the turmoil among some of the world's nations, which is seemingly never ending, it is likely to stay on the agenda for some time to come. The first six of the 17 UN Goals, all of which have links to poverty, include food/hunger, health, education, water, sanitation and energy.

There is the addition, recently, of very probable human-induced changes to the climate leading to alterations to the environment, some significant. These are seen as largely a result of a very large increase in the population numbers over the past 60 years (see Figure 1.1). This increase in population, in turn, is partly due to positive results from medical research put into practice, improving life expectancy, and partly as a necessary result of enhanced food production and quality.

3.3.2 Issues Underlying Poverty

Seemingly endless wars and conflicts have caused an upsurge in the number of people moving, including refugees, looking for new and better locations to find peace and probably overcome poverty in some places, as well.

However, the other reason for people to move is insecurity, apart from war, and even danger brought on by the type of government regime acting in their particular part of the world and putting constraints on their people's activities. Apart from the environmental factors, it is the manner by which people are governed, the freedom they have for opportunity and self-expression, that decides some of the people, who have the ability and opportunity, to move onto another country and seek a better and more peaceful life.

Today, there are large movements, in the order of hundreds of thousands of down-pressed refugees in temporary camps providing shelter, for example, in Sudan, as a result of civil unrest, sometimes war with tribal and religious overtones; in Somalia, as a result of war-like religious fanatics; or Myanmar, the result of an oppressive military dictatorship which took over the country in a coup from an elected civilian government, and more. In the Americas, there is a continual movement, from south to north to the USA and Canada, of people moving away from the control of drug lords or government corruption. In Africa, people are moving from unacceptable levels of pervasive poverty, poor governance and government corruption to Europe, seeking a better life with the apparent freedoms the northern countries can offer.

Then there are countries which are simply too poor, either because of an excess of population over resources or insufficient resources to support the existing population, and lack of education. In East Africa, the island of Madagascar had a population expansion over the past 60 years from 5 million to 28 million, unwittingly destroying the forested environment to create more much-needed agricultural land. In the process, the change in land use altered rainfall and erosion patterns significantly, and instead, the people unwittingly brought about drought and crop failures, much like the Easter Islanders in the Pacific did a few centuries ago. Bangladesh, formed by the washdown sediment of the major Ganges and Brahmaputra River systems, despite showing recent modest economic growth, is a relatively poor country. The country suffers from having to support a population larger than its natural resources are comfortably capable of. Uganda and Kenya have shown significant impact from population growth figures of approximately 7 times in the 60 years since 1960 and decolonisation.

There are many organisations and charities, not forgetting the UN, as well as the governments of some of the richer nations, which provide much funding and are working in many areas around the world to provide much needed assistance. Without this help and the sustenance, which is generally brought with it, the populations of the world would in many places have shown signs of collapse.

The many reasons why there is poverty means that there have to be different solutions found. In some cases, the problem is so vast that there is little more that can be done with the resources that can be made available, but to maintain the status quo. In other cases, however, there is an opportunity to bring the people out of poverty and help them to the point at which they can help themselves as a community. In order to be long-term successful, the people will need to have developed suitable skills, means of earning some returns, for example, from agriculture or fishing and its support functions, hopefully more than just at a subsistence level, that they may prosper.

Apart from Yemen, for example, which has been under great duress for the past 8 years from religious-politically motivated conflicts, the continent of Africa is arguably where the greatest areas of poverty arise.

There is an acute shortage of water in these places, not just clean as desired, but in total terms. In addition, there is great hunger because of food shortages, sometimes famines, mainly because of poor ground, continual heat and lack of rainwater. Consequently, there are starving communities in their thousands who appear frequently in the media. The problem of poverty is quite often likened or connected to corruption, where there is, too often, a government of self-serving individuals.

The population of Africa is currently 1.42 billion, and this figure is expected to rise towards 4.2 billion by the end of the century. By that time the continent of Africa will host some 37% of the then world population, compared with today's approximately 17.5%. The growth rate annually is at 4.5%, which is effectively unsustainable, particularly in the conditions that prevail. For instance, electricity services have a long way to go to meet demand, and power cuts are frequent. From the growth point of view, there are thousands of children being born with inadequate sustenance in the important first days of life who will grow up with lives which will be stunted, if not foreshortened. This, from a humanitarian point of view, is truly unacceptable, and feeding newborn and young children is probably one of the most prominent tasks within the remit of the UN's Sustainable Development Goals.

The UN has declared Yemen as a humanitarian disaster area, with many children dying from malnourishment. It is the result of interpretations within a religion adding to a secular situation. The situation has become even more complicated latterly, because of the Israel–Palestine conflagration. The Yemen-based Houthis, in support of the unfortunate and battered Palestinians, are attacking shipping using the Red Sea, such that several international lines have diverted and are taking the longer route, with added costs, round southern Africa and the Cape of Good Hope.

In the sections which follow, examples are given from working examples of projects, one dedicated to alleviating poverty in East Bali, Indonesia and the other to a range of projects across Pakistan over a 25-year period which have been mainly directed to dealing with poverty or the consequences of natural disasters. Reference is also made to the December 2004 Aceh tsunami disaster, which was the largest natural catastrophe in the past 100 years with a total of some 286,000 lives lost.

3.4 EDUCATION FOR A POVERTY PROJECT

3.4.1 Project Background

The benefits of providing good education were brought home to those of us who had had the privilege of having had a full education, when attending to a poverty situation, namely the East Bali Poverty Project (EBPP), over the past two and a half decades (Younger et al. 2012). The location of the project is on the northern shoulders of Mts Agung and Abang in Bali. Mt Agung last erupted fairly violently in 1963, impacting the forefathers of today's people, and reminding us again by dangerously rumbling recently in 2022. Many survivors of those affected in 1963 were forced to go away to South Sumatra, an estimated 50,000, with whatever livelihood they had, because the land was covered in ash spewing out from the volcano and critically affecting habitability. This was one of the largest volcanic eruptions of the 20th century.

When the project was started in 1998 by the founder, David Booth, a professional civil engineer with extensive overseas experience, the number of residents was 17,000, but their status and future outlook were grim and gloomy. There was 85% illiteracy in the community, the infant mortality rate was significantly higher than world norms, the average longevity of the people was barely 40, and there was an air of dejection. They lived in 19 different villages/hamlets, rarely mixing, within a district (*desa*) called Ban, which is shown in Figure 3.1, and six of the villages were so remote that access to a government school was at least 4 hours away over steep and winding mountain tracks. No public transport was available, even to this day. This discussion, nevertheless, is concentrating first on the education aspect of the project; the other components of the project,

FIGURE 3.1 EBPP location.

which, naturally, were vital and had to be dealt with at the same time, are covered in the section that follows, dealing with poverty.

3.4.2 Project Initiation

Before the project got underway, David Booth, on behalf of EBPP, signed an MoU with the community leaders and the children's parents, whereby he made three promises and the parents also made three promises. David Booth promised to provide integrated and relevant education so that they could lead their communities into the future; secondly, to provide a nutritious meal every day with a glass of fortified milk and a multivitamin tablet and thirdly, to teach them how to stay healthy without medicines. The parents' promises were to send the children to school every school day, provide space, support for them to study at home and lastly, learn from their children.

However, at the beginning, the first few months were taken up with the founder of the project, EBPP, David Booth, getting to know the people, getting their trust and learning what were their problems and which were the most urgent, that is, water and education.

Initially, the people were getting to know David. In brief, the people said they wanted their children to have a better life than they had had, be educated and have lifelong opportunities. David knew he was effectively starting from scratch, and he laid down some ground rules, perhaps most important being 'no corruption', for which he received a pleasing wholehearted agreement. Then, some local staff were recruited, and the two main are still engaged to this day.

With only modest funds at its disposal, the poverty project was started with a small but purposeful office located just to the north of the main tourist centre of Bali. A survey was made of the condition, health, and educational ability of the people. These were among the first tasks, apart from attracting the supporting funds for the project.

It was found, as said above, that there was a shockingly bad 85% illiteracy level, including most of the adults; the infant mortality was significantly above world or even national standards; and the average age for life was below what Indonesia had achieved in the 30 years from 1970. Obviously, these people had been 'left behind' within the impressive improvements across Indonesia over the same period.

3.4.3 Project Activities

Access to and communications at the start were decidedly primitive, and sorting these out was a priority as well as finding a quick method to get access to water and, where possible, to direct the water into the different villages.

Six schools were built across Desa Ban, and in a comparatively short time, all children, including those in the six most remote hamlets, benefitted from free elementary education with most continuing to Junior High School level.

The first school was ready at Bunga (shown in Figure 3.2), but the desks were still awaited. It did not stop the first intake of youngsters from attending and just sitting on

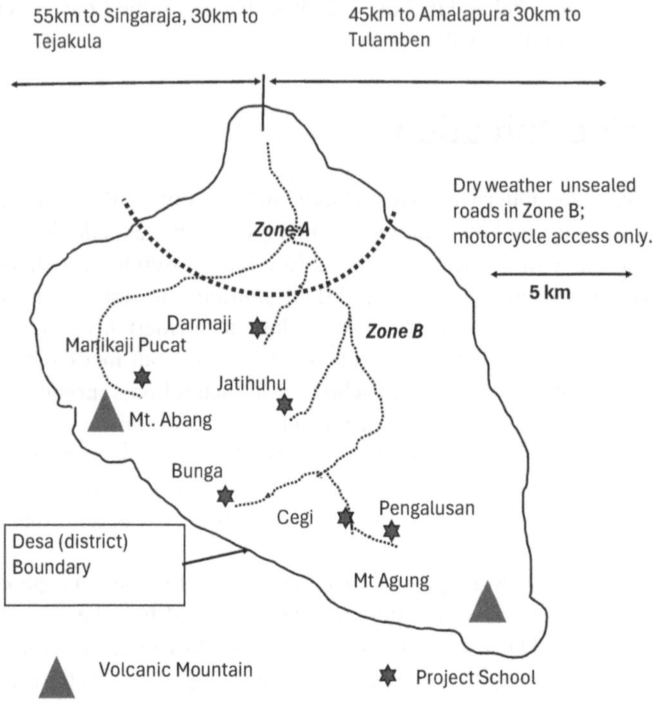

FIGURE 3.2 School locations.

the floor. Fortunately, desks made locally by the parents very quickly followed. The air of excitement was palpable when they received pencils and paper, and humbling for David and those of us present to see. The children eagerly awaited appropriate books of learning, and a library gradually built up. The schools' programmes run from primary through to senior high school and match the government terms and curriculum with the additional attention to the environment, particularly in which they reside. They have been taught to respect the environment. As found elsewhere in the world, however, there was a noticeable drop off in attendance towards the end of junior secondary. Sometimes due to pressures from home.

The other five schools were located at the most remote of the hamlets of Desa Ban, namely Cegi, completed in 2000, Pengalusan in 2000, Manikaji in 2001, Jatituhu in 2005 and Darmaji in 2007. Procedures were set up to start with for the Bunga school and, following their success, were replicated at the launch of the other five schools and were subsequently adhered to through until today.

The different phases of the project were introduced, starting from sensitively designed model programmes to address the early health issues, the education programmes, road access, appropriate and sustainable organic vegetable and herb farming, home-based rainwater harvesting and the installation of solar panels. All programmes have a common underlying theme, that of garnering human and natural resources harmoniously for sustainable common benefit. The research-related costs, which started once the project research centre was set up, helped by a donation from Unilever in 2004, underwrote the practical development themes that were designed to benefit the project.

These themes included training programmes on topics such as library systems, trials on vetiver and other plants acceptable to the environment, tree species and growing bamboo, which could provide a future livelihood for some of the population. Once the project was deemed to have reached a reasonable stage of sustainability, it focused on activities to provide sustainable food security and livelihood opportunities in the area, which could be replicated in other similar projects.

An early task was upgrading the access. Initially, all roads were dirt tracks with no vehicular access till the first road to the area was built from the saddle between Mts Agung and Abang. This access to Desa Ban was chosen particularly to connect to the remote communities, and the location to start the education for the children. The road access was delivered by members of the communities themselves, after receiving instruction and guidance towards a simple design for a concrete surfacing, with the cement provided by the project.

The first road, as indicated, to the village of Daya was constructed from the saddle between Mts Agung and Abang. This provided the first access in history for thousands of people to leave the villages and for food, medical and other supplies to be able to enter, and to facilitate learning. To date, some 25 km of road infrastructure has been facilitated by the project linking most of the hamlets and maintained by the respective communities, as required. The other main activity at the start concerned access to water, to overcome the need to walk up to 5 hours over very difficult terrain to fetch it.

As of mid-2006, the project had facilitated access to safe water supplies for 2,400 families initially from mountain springs, courtesy of Thames Water International's annual charity donation, and also using artesian wells and/or *cubang*s, and roof rainwater collection reservoirs built by the communities. A typical example of a *cubang* is shown in Figure 3.3. Virtually all residents can now access safe drinking water.

FIGURE 3.3 A *cubang*.

Work on sanitation facilities was started in 2010, and 1,515 families had facilities in the communities by 2022. All families were catered for in extended family compounds, when facilities were not for individual families. With materials and guidance from the project's water and sanitation team, individual toilet/bathroom blocks with septic tanks were built by the respective families, community by community, as project funding became available. In 2010, to start this very important part of the project, Royal Bank of Scotland (RBS) provided funds from their annual charity donation, and the team were pleased to welcome Sir Jackie Stewart, former Formula 1 world champion, in his ambassadorial role on behalf of RBS.

Another vital element of the work completed by the EBPP team concerned the health of the communities. In this, the families benefitted from attendance at 27 *posyandu* (monthly mother-infant health posts) initiated by the project. These were run locally by volunteer local cadres, trained by EBPP, and the project was sponsored by UNICEF in 2003, and did not exist prior to 2003. Since attending these, the infant mortality rate has been reduced from 30% to zero. Dental healthcare was facilitated by the project, accessing most schools by trial motorcycle or 4×4 vehicle, for all children from 2002 until 2020, when complete dental facilities and treatment became available at the nearest government community health centre *(Puskesmas)* in the nearby district of Tianyar.

Physiotherapy has been given to 110 children who were born with disabilities. Cleft palates, cataracts, and other eye disorders have been attended to for 91 children. Three hundred adults have been treated as well. Ear, nose, and throat clinics have been provided for all school children annually since 2011.

As part of health, food and nutrition are watched carefully for the communities, since there are families which are not able to cope adequately, as is found for most of society in general. In the course of the work, by 2023, the project had supported 111 severely malnourished and 244 moderately malnourished children. Iodine deficiency disorder, which stood at 84.5% in 1998, was completely eliminated.

Returning to the roads, small sections of other tracks had to be upgraded, and these received attention as funds became available. As several sections of road have to cross steep terrain, a safe and sound method in line with the ground conditions had to be considered to stabilise the slopes. It was found that vetiver grass (*Chrysopogon zizanioides*) was ideally suited for the task. It grew fast, reached 3 m high but, more importantly, the roots also penetrated quickly and could be up to 6m deep and have a tensile strength one-sixth that of mild steel. It had been acknowledged by the World Bank in 1987, and also Booth and Adinata (2003). Since then, the Bali project has been assigned to give its expertise around the region for other appropriate projects. Figure 3.4 shows the use of vetiver grass on the project.

Bamboo is the other grass which is utilised on the project. It is grown and used by all the communities where it plays a major role in Hindu Bali community cultural and religious ceremonies. It is fast growing and strong, often being used as a structural member, for example as formwork, in locations across the world, where it is available. A total of 80,000 bamboo poles of various species have been planted on the project in suitable areas, in collaboration with the communities, covering a collective total of some 200 Ha, over the decade from 2007. Some of the bamboo has been used to make products for the project's 'East Bali Bamboo Bikes' social enterprises for the frames of bicycles, and many different types of handicrafts (Booth et al., 2019).

FIGURE 3.4 Vetiver grass by strengthening roadside earthworks.

FIGURE 3.5 Installing new solar panel at Bunga school.

The first power for the project was provided by solar panels put in from 2007 to 2009, as funding became available, for the six community schools, enabling teachers to use computers and projectors and significantly widen the potential horizons for the children. Solar power was very rare for that time in Indonesia, despite the fact that Indonesia has significant reserves of solar power (Younger et al., 2022). It was recognised that it would be some considerable time before any public services reached the area from government. However, the government followed the project's lead and had fully provided power from a massive solar farm located in Kubu on the northeast coast, enabling all schools, households, etc. in Desa Ban to be connected to the national grid by 2016 (Figure 3.5).

Other possibilities, such as wind and micro-hydro, were investigated, but were not found to be realistic for the area.

Research in 1998 had identified deep wells that had been installed in the late 1970s in a dry riverbed, supplying the only dry season water for the 400+ families of the three hamlets of Jatituhu, Darmaji and Temakung. The project, with the full participation of the three communities, cleaned, repaired and sanitised the three wells and eventually installed solar-powered Lorentz water pumps. The completed installation of the pumps, providing up to 23,000 l/day from the three connected wells, without the need for an inverter. The Lorentz unit, which can pump up to a vertical height of 270 m, automatically shuts off when the water tank is full. Maintenance is minimal and life expectancy is 10–20 years. Both the pumping system and solar arrays are earth grounded from damage due to thunderstorm-induced electricity.

One interesting experiment was with the *neem* tree that has multiple benefits. In this relatively dry land situation, its tap roots go down 15–20 m into the ground, which can draw water up for its own growth and help to lift the level of the water table for future land fertility.

As the project progressed, there was obviously a need to change the emphasis of the things that had to be done on a routine basis, but not forgetting those that had to be handled with a certain degree of regularity and rigour, for instance, education and health.

There are many lessons that have been learnt from EBPP and that can be copied elsewhere. This is a private sector initiative from the passion of one individual. Many humanitarian aid agencies are responsible or are known for providing a major vital need, for example, water, which is in short supply in several, too many countries of the world.

Within each of these elements that have had to be tackled by EBPP, most attend to poverty, as well as the environment. Sometimes the numbers become millions, and if they are largely indigenous people, and not refugees driven by other factors and wanting to move on, they must be helped to stay put and learn to develop sensibly the environment in which they find themselves. A significant task is to ensure that communities become literate which, in turn, gives confidence.

It has been found that smaller groupings are manageable, much more sustainable, just as developed countries have found when teaching class sizes. It has certainly been noticeable on the EBPP.

3.4.4 Carrying the Experience Forward

The EBPP experience can be replicated around the world, but it is well understood that the situations vary greatly elsewhere and there are many hundreds of thousands to deal with when there are 1.6 billion at the poverty level, a monumental task as the UN knows well. The EBPP has indicated the success of having the whole community involved and benefiting from that involvement. If there is a lesson from this experience, it is that the groupings of people should be at community level to forge the best results.

It is strongly believed that children's education is a fundamental human right and a foundation of sustainable development and poverty alleviation. Six schools were built by members of the communities themselves with suitable guidance from staff, and comprehensive education programmes delivered in the most remote mountain communities in partnership with all stakeholders. Work was carried out with community leaders and parents to encourage and support the children to continue learning till high school graduation so that they could have a better future and, at the same time, help develop their impoverished communities with new production and fresh ideas, and also to extend their ability to meet the challenges that they would meet beyond Desa Ban.

One of the delightful findings was the apparent latent talent for art which showed up in the children's paintings. This was part of a wide range of extracurricular subjects. The children were being empowered with life skills and the capacity building they needed to forge their own futures, even by starting their own businesses in their community, or continuing to university, or, thirdly, seeking good work opportunities beyond Desa Ban.

With regards to tertiary education, and in line with EBPP and community aspirations, sponsors were sought for the graduating students who had the potential and enthusiasm to continue to university with the goal of returning to Ban to share their new knowledge, and participate in developing a new generation and leading the village people to a better future and more sustainable social and economic development.

It is worth repeating the testimony of a former student, now a mathematics teacher for the community. "There was no education available in my area until I was six years old, the time when EBPP opened their school in my village, Manikaji, in 2001. Today I am one of EBPP's teachers and I am dedicated to make sure the children in our village continue to receive the best education" (Nyoman Kartya, Mathematics Teacher).

To date, there have been 1,200 receiving education; 450 have graduated primary school, 234 at junior high level, and 107 senior high level. Twenty-one have earned scholarships to go to university, and it is believed this trend will continue as the children who follow see the successes of those who have preceded them.

The experience of EBPP in the last quarter of a century can be deemed a success, and EBPP has modified its role in terms of what the people can now do themselves. The communities in the district no longer feel isolated and the younger generation have developed a confidence to handle most things that come their way. The role that EBPP takes now is one of handling outstanding matters, a watching brief, any maintenance of physical assets required and continuing the fund-raising for these as required. It is also worth remembering that for a successful community it needs education, but it also needs regular sustenance and peaceful surroundings.

A point from this is the size of the project area and the community numbers. It is a comparatively small area, and the numbers are also relatively small compared with whole countries, or even part of any country that is the aim of one or more charities or the aid arm of the UN or one of the donor nations. This disparity is discussed later, but first the East Bali project and the possibly different comprehensive nature of it, with the emphasis on 'comprehensive', is covered, as well as the lessons that could be taken from it. Latterly, the project has been recognised as one of the ten best NGOs in Indonesia.

Further and some fuller details of the project are given in the *Proceedings of the Institution of Civil Engineers, London* (Younger et al., 2012, 2017).

3.5 HELPING RURAL COMMUNITIES IN PAKISTAN

3.5.1 Background

In the past three decades, Azra and Peter Meadows, working from Glasgow University, carried out a number of projects with a significant sustainable and environmental impact, including the problems related to poverty. Following Peter's death in 2023, Azra continued this work.

The aim was towards uplifting rural communities across Pakistan with the help of a wide range of sponsors. Geographically, the work covers a wide range of communities in coastal, riverine and mountain regions and the plains across the country, as can be gauged from the discussions which follow, based on key references listed in Table 3.1.

TABLE 3.1 List of key references by A. Meadows and P. Meadows

YEAR	*TITLE*	*SOURCE*
1990	Mountains, rivers and the coastal zone in Asia: environmental management into the 21st century	Meadows P.S. and Meadows A.(1990)
1999	The Indus River: biodiversity, resources, humankind	Meadows A, and Meadows P.S.(1999)
2003a	Environmental sustainability and development of mountain rural communities in Pakistan	Meadows A. (2003a)
2003b	Natural hazards in mountain areas	Meadows P.S. (2003b)
2004a	Interaction between rural communities and mountain diversity	Meadows A. (2004a)
2004b	Ecosystem biodiversity. Environmental stress and natural hazards in mountains	Meadows P.S (2004b)
2011	Wildlife conservation education and international programmes	Ali Z., Meadows, A, Meadows, P.S. et al (2011)
2011	Ecosystem sustainability, climate change and rural communities	Meadows P. S. (2011)

3.5.2 Indus River: Biodiversity, Resources, Demographic History

The work started with the organisation of an all-embracing symposium dealing with the Indus River, at the Linnean Society in London in 1994. The symposium covered the extraordinary range of ecosystems, biodiversity, mineral resources and demographic history of the unique Indus River and surrounding regions in around 2600 BCE. It included highlighting the ancient city of Mohenjo Daro, which with its paved streets laid out in a regular pattern, had been very advanced for its time. However, after seven centuries of prominence, it was impacted by the Indus River changing course, causing the city to go into decline. The city population scattered widely and sought new pastures, many going westwards and ending up as early European settlers. Mohenjo Daro is still being examined (Meadows and Meadows, 1994).

These discussions enabled future scenarios on enhanced resource development to be evaluated while protecting the unique diversity of the range of ecosystems along the river's length.

3.5.3 International Hindu Kush Expedition

The second undertaking by the Meadows at Glasgow University with the staff of the University of Peshawar was to the Hindu Kush, a 2-month long multidisciplinary international expedition to Chitral, involving 40 people from UK, USA and Pakistan. The objectives were to study the impacts of mountain rivers on rural communities and the environment in Chitral, Hindu Kush. The expedition concentrated on the narrow river valleys of Garamchashma, Bumburet and Bareer (the people who dwell in this village, the Kalesh, are descendants of Alexander the Great's entourage and have a unique tradition from that time) and Rumber and the broad river valleys of Chitral and Mastuj. The study was divided into four areas, namely the river function, natural hazards and biodiversity, secondly the present-day human settlements, thirdly prehistoric and historic human settlements and lastly the geomorphology – landform and use. Some of the intended outputs of the expedition were training of Pakistani members in state-of-the-art methodologies used by overseas members. Further, it involved learning by overseas members of the problems of working in isolated and hazardous environments and increasing the awareness of local communities to the mechanisms, consequences and avoidance of natural disasters.

At a meeting, held at the Linnean Society in London at the end of 2000, ongoing explorations in Chitral, and some initial findings of this expedition were discussed. This was followed by a 3-day conference in 2001 at the University of Peshawar, in which international and national delegates delivered talks on the results and outputs from the expedition.

3.5.4 Higher Educational Programmes

In parallel, were Higher Educational Link Programmes which were funded by the UK government (Department for International Development) and managed by the British

Council. The purpose of these was the alleviation of poverty and suffering, to improve quality of life and enhance sustainable development, and empowerment of women in developing countries.

3.5.5 Coastal Zone Management and Environmental Impact Assessment (EIA) in the Mangrove Swamps, Indus River

Over a 4-year period a project was undertaken between Glasgow University and Karachi University, to look at the state of the Coastal Zone Management (CZM), very important to the country's well-being. Several field trips to coastal villages highlighted the lack of safe fishing boats. Community work with the World Wildlife Fund staff to enhance mangrove reforestation proved essential. Eight academic staff from Karachi University underwent training in research methods, IT transfer and literature searches at the University of Glasgow. A 3-day national workshop on CZM and EIA was organised at the British Council in Karachi which resulted in several important recommendations for the coastal fishing community. In addition, a 2-day training workshop was organised for students and staff at Karachi University on 'IT Transfer: Statistical Analysis of Marine Science Data', which enhanced their data analysis skills.

In 2001, a small expedition was undertaken along the Makran coast towards the border with Iran, visiting five areas, Pasni, Astolla Island, Gwadar, Ganz and Jiwani. This trip assessed the pristine coastline and its marine resources and the livelihood of the fishing communities. In the remote village of Ganz, it was possible to conduct a discussion with the villagers who, after long reflection, highlighted their top priority was a jetty, so they could safely anchor their boats, as they totally depended on their catches for their livelihood.

3.5.6 Socio-environmental Uplift of Rural Communities in Sindh and Baluchistan

Following this successful study, Glasgow and Karachi Universities undertook a 3-year project (2003–2006) dealing with the socio-environmental uplift of poor rural communities in Sindh and Baluchistan. This project was cross-disciplinary, partnering with social scientists, aiming to enhance the quality of life of poor coastal communities, by recommending practices for use of safe drinking water and hygiene. A female community workshop was organised to teach them ways of treating water to make it safe for drinking. A pamphlet was published in Urdu with diagrams illustrating basic methods of treating water to make it safe for consumption. Sociological surveys were also conducted in the villages to assess the drinking water quality of these deprived communities. A half-day seminar was organised at Karachi University on 'Safe drinking water for better health' with talks on the importance of this issue and ways to improve clean water availability for the deprived coastal communities.

3.5.7 Sindh Coastal Community Development

This was an ADB-funded 6-year programme (2007–2013). The Glasgow team were invited, as consultants, by the ADB to advise on the biological and environmental aspects of the project, and to assess the present state of the environment and poverty levels in rural coastal communities of districts Thatta and Badin in Sindh Province. In addition, recommendations were made for improving community livelihoods and environmental sustainability. The assessment was conducted in 2006 and resulted in recommendations to ameliorate village poverty, improve institutional capacity, and promote sustainable use of the unique ecosystems of the region. These proposed interventions for planting strands of mangrove trees, upgrading marine and freshwater fish hatcheries, developing aquaculture ponds for fin fish and establishing systems for the culture of shellfish, seaweeds, and salt-tolerant plants.

3.5.8 Development Partnerships in Higher Education

At the same time (2007–2010), the University of Glasgow formed a very productive link with the University of Veterinary and Animal Science at Lahore. In this 3-year project, a good number of participants attended training courses which were wrapped up in a conference. The rural community members benefitted from the community workshops and interaction with staff and students. Three village visits were conducted with village community meetings, and a River Indus boat trip was taken to Taunsa Wildlife Sanctuary.

This was followed by a discussion and Q/A session about the livelihood and development needs of the communities. The situation and needs were found to be as follows:

1. **Health**: There were no maternity facilities or dispensary, with the hospital 5 km distance.
2. **Transport**: Only public wagons were available.
3. **Treatment**: Costs high – hepatitis B, tuberculosis, malaria, in particular.
4. **Water**: Safe drinking water was needed, as well as sanitation.
5. **Education**: This was inadequate for 200 households, served by only one male and one female teacher on loan with one classroom and no furniture.
6. **Fishing (Fisheries and Wildlife Department)**: Current arrangements were not at all helpful; corruption present, pesticides being used to kill fish and turtles; vocational training urgently needed in water conservation and correct fishing methods.
7. **Women's opportunities**: There was an urgent need for resource training centres for promoting handicraft skills.
8. **Others**: There was too much unemployment as well as bonded labour, which spirals debts, mistreatment of villagers by contractors and an absence of veterinarians to tend to livestock.

The villagers were unsure of their rights, and one village was facing the problem of being asked to leave their land, as they came and settled at the time of the partition,

1948, without legal allocation to the land they occupied. There is not a healthy working relationship with local landlords. (Similar problems to this day from the same period of end of British jurisdiction in Palestine, partition of the land in 1948 to allow the creation of Israel.) The labour rights are also not there. This shows a lack of awareness. There is a need to implement the villagers' rights and give them proper guidelines. Poor governance by officials has led to the revenue not being spent on villagers' health, education and social welfare issues but rather on the government officials' domestic matters. This situation is sadly replicated in many countries around the world, whether in Africa, Asia or South America, and is a significant factor in holding back development.

The University of Glasgow and core staff undertook a boat survey along the River Indus to assess its biodiversity and its habitat status. The variety of wildlife was unique and included the endangered Indus River dolphin, wildfowl, turtles and the vegetation. These activities indicated how the rural communities utilise the natural resources for their sustenance and livelihoods.

3.6 DISASTER MANAGEMENT AND HUMANITARIAN ASSISTANCE

3.6.1 Scottish Government International Development Fund, South Asia Programme

In 2005, northern Pakistan suffered one of the most devastating earthquakes ever recorded, followed by two countrywide floods in 2010 and 2011. These catastrophic events were deemed to have set developments back by a decade. The significant loss of life and destruction caused serious food shortages and suffering. Along with the Heritage Foundation of Pakistan, Glasgow University, with the aid of two Scottish Government grants, assisted communities in Khyber Pakhtunkhwa Province to enable them to restart their lives with the aim of expanding the work on strategies for early recovery and sustainable development in post-disaster and post-conflict communities.

3.6.2 Emergency Shelters for Flood-Affected Communities: Swat Valley

This was a 6-month project to build emergency homes for families displaced by the unprecedented 2010 floods; 266 new homes were built in the Swat valley. These were units built with sustainable local materials – bamboo, stone, lime-mud – on a guided self-help basis, and would last until permanent homes were built (*Note*: the earlier discussion from the EBPP).

3.6.3 Disaster Rehabilitation for Women, Siran Valley

This 3-year project (2010–2013) provided disaster management and rehabilitation guidance for rural communities affected by the 2005 devastating earthquake in the Siran valley in the northern region. The programmes aimed to educate women and children about natural disaster preparedness and post-disaster actions, and on assisting families to develop handicraft skills and livelihood programmes. These included provision of clean water, community food programmes and novel-based agriculture. It also provided advice to the menfolk. One of the major outputs was the publication of an illustrated *Disaster Preparedness Manual* (2013 in English with Urdu translation) and an associated DVD.

The two SGIDF projects in Khyber Pakhtunkhwa Province have been so successful that they are now being replicated and implemented in the coastal zone of Sindh Province with the involvement of local rural communities.

3.7 POVERTY AND DISASTER: THE ACEH EXAMPLE

Before the tsunami struck the west coast of Aceh Province on 26th December 2004, Aceh ranked as one of the poorest provinces of Indonesia with poverty widely present. The disaster is probably the largest in the world this century, triggered by a very strong offshore earthquake, 9.0 on the Richter scale, and accounting for 286,000 lives. These were lost in Aceh and across the Indian ocean, although mostly in Aceh with the town of Banda Aceh, lying at the north-west point, bearing the brunt. Phuket in Thailand and the east coast of Sri Lanka were also badly impacted.

In addition, Aceh province was also troubled by a conflict, which had been going on for some years involving an Islamist separatist group, Islamic Separatist Group (GAM), in 'control' of the central hills of the province, sparring with the government's military, the Tentara Nasional Indonesia (Indonesian Armed Forces). Because of this, the province did not attract inward investment even from national budgets, hence the high level of poverty, although the province held good supplies of oil and gas and was well provided for in other natural resources. The tsunami event changed everything.

An article in 2012 (Younger, 2012) shows the extent of the relief work undertaken by many people to reinstate the various parts of the town to make Banda Aceh alive once more.

What is more, GAM was brought into the fold, and forms part of the provincial government with sharia law allowed, under special dispensation from central government. The province (the most Islamic of all the Indonesian provinces) is once again going forward. In this unusual story, an awakening from a seemingly never-ending dispute, as a result of a disaster, brought the parties together; their quarrel deemed being much smaller than the urgent needs of the people in the province as a result of the disaster. Regrettably, there are other places in the world which could follow this example.

An interesting sideline latterly is that the province has accepted exiled Rohingya refugees, escaping from the brutal conditions, sometimes deadly, imposed by the Myanmar military.

3.8 CONCLUSIONS

As Brown (2021) has argued, having had nearly a decade of serving as an UN ambassador and travelling to many places, education is critical towards alleviating poverty. One is talking about attending to a major problem that invokes many hundreds of thousands of children mired in poverty and worse. There is no doubt that dealing with hunger and thirst, healthy living and security must be seen as an imperative to be provided at the same time, if children are to benefit from learning.

In summary, in the nearly three decades that the Meadows and Glasgow University staff have been involved in a wide range of projects in Pakistan, including response to a major disaster, an improvement towards the overall level of poverty was witnessed. Unlike the East Bali project, in Indonesia, over much the same period of time, which was solely dedicated to a basically forgotten community of people at the bottom level of society and showed how poverty could be alleviated with great success and the community flourish, the work in Pakistan was carried out in more traditional ways. In this, key aspects of the different communities' problems were targeted, for example the fishing folk of Sindh Province, who lacked the proper facilities, or overcoming the very real problem of beggars and transforming their livelihoods at Makli. As indicated previously, a given poverty problem has to be scrutinised, and the appropriate method chosen and action taken to help alleviate it.

Following an opening to the subject in which poverty was discussed in a general sense, it was maintained that poverty would continue to be around affecting humanity in many places in the world for the foreseeable future, negatively impacting on sustainable development. Examples have been given in which a wide range of solutions was described to tackle the different problems arising.

In virtually all cases where poverty was found, there was a solution, not just needing money but most of all the right help and attitudes to tackle a given problem, with self-help with confidence instilled being the ultimate aim. There remain areas of the world in Africa and South America, as well, where there are never-ending non-governable problems. There is a self-serving elite, supported by drug trafficking, in S. America for instance, or extreme religious intolerance holding sway in parts of Asia and north-central Africa, and corruption present throughout in most places.

The numbers showing for abject poverty, 1.6 billion, in the world have not altered much in the past 60 odd years, although the overall population has shown unprecedented growth from 3 billion to over 8 billion today. This has largely been due, as said, to the better health from the great strides that have been made in science and medicine which have reduced infant mortality, and allowed people to live healthier lives and significantly longer. One may debate the issue whether the extent of the advances would have occurred, had not the advent of the computer taken place and accelerated at great pace that we are now in the Digital Era, with virtual reality and AI now beckoning. Probably not!

Looking ahead, we have the understanding and the ways and means to reduce if not eliminate poverty entirely. Perhaps, like corruption, we aim to curb it as much as possible and recognise and take action wherever it is found. However, unlike corruption, poverty can be eradicated entirely in given areas or situations, and we must strive to widen our ambitions. The resistance to elimination of poverty in some situations is authoritarian, selfish, often brutal rule which cares for its self-preservation above all else. There are still too many countries which exhibit that tendency, which, alas, hold back poverty alleviation and its part in overall sustainable development.

However, the opportunities that humankind have opened up going forward are highly significant and unprecedented compared with just a century or even half a century ago. Living on the Moon or Mars! (if the desire is there!) But there is also a duty to the less fortunate in society. We must not forget, and the skills that are developed for countering poverty will still be required for the foreseeable future.

REFERENCES

Ali, Z., Meadows, A., Meadows, P. S., Qureshi, N. A. and Butler, J. R. A. (2011) Wildlife conservation and rural livelihoods in Pakistan. Future challenges and opportunities in a globalising world. *Journal of Animal and Plant Sciences*, 21(2 Suppl.):441–442. https://www.thejaps.org.pk/docs/21-2-suppl/JAPS-21(Final).pdf

Booth, D. J. and Adinata, N. A. (2003) *Vetiver Improving Lives of Impoverished Indonesian Subsistence farming Mountain Communities, Led by Children*. Available at: https://vetiver-indonesia.wordpress.com/wp-content/uploads/2012/02/ok-vietnamconf.pdf

Booth, D. J., Younger, J. S., Parry, D. E. and Suastika, K. (2019) Applications of the construction grasses, bamboo and vetiver on the East Bali Poverty Project. In *Construction Conference*, Bandung, Indonesia, April.

Brown, G. (2021) *Seven Ways to Change the World*. Simon and Schuster, London.

Kirwan, C. G. (2024) *Smart Cities: Artificial Intelligence*. Elsevier, The Netherlands.

Lari, Y., Meadows, A. and Meadows, P. S. (2013) *Disaster Preparedness Manual*. Text in English and Urdu. Heritage Foundation, Karachi.

Macaskill, W. M. (2022) *What We Owe the Future*. One World Publications, London.

Meadows, A. (2003a) Environmental sustainability and development of mountain rural communities in Pakistan. Publ in *Mountains of Pakistan Protection, Potential and Prospects*. Eds. Mufti, S. A., Hussain, S. S. and Khan, A. M., Global Change Impact Studies Centre, Islamabad, Pakistan, 133–152.

Meadows, P.S. (2003b). Natural hazards in mountain areas. In *Mountains of Pakistan; protection, potential and prospects*. Eds. Mufti, S.A., Hussain, S. S. & Khan, A. M. Global Change Impact Studies Centre, Islamabad, 100–117.

Meadows, A. (2004a) Interaction between rural communities and mountain diversity. In *Biodiversity of Northern Areas of Pakistan*, Ed. Bano, A., Islamabad. 209–228.

Meadows, P.S. (2004b). Ecosystem Biodiversity. Environmental Stress and Natural Hazards in Mountains. Publ. in *Biodiversity of Northern Areas of Pakistan*. Ed Bano, A. Printed in Islamabad. pp 262–283.

Meadows, A. (2011a) Wildlife conservation education and international programmes. *Journal of Animal and Plant Sciences*, 21(2 Suppl.):305–316. https://www.thejaps.org.pk/docs/21-2-suppl/JAPS-1(Final).pdf

Meadows, A. and Meadows, P. S. (1990b) Mountains, rivers and the coastal zone in Asia: Environmental management into the 21st century. In *Integrated Environmental Management, Development, Information and Education in the Asia-Pacific Region*. Shiga University Organising Committee, CRC Press, Lewis Publishers, 49–76.

Meadows, A and Meadows, P. S. (1994) *The Indus River: Biodiversity, Resources, Humankind.* Proc. Symposium published by Linnean Society of London.

Meadows, A. and Meadows, P. S. Ed. (1999) *The Indus River: Biodiversity, Resources, Humankind.* Oxford University Press, Linnean Society, London.

Meadows, P. S. (2003b) Natural hazards in mountain areas. In *Mountains of Pakistan; Protection, Potential and Prospects*. Eds. Mufii, S. A., Hussain, S. S. and Khan, A. M., Global Change Impact Studies Centre, Islamabad, 100–117.

Meadows, P. S. (2004b) Ecosystem biodiversity. environmental stress and natural hazards in mountains. Publ. in *Biodiversity of Northern Areas of Pakistan*. Ed. Bano, A., 262–283.

Meadows, P. S. (2011) Ecosystem sustainability, climate change and rural communities. *Journal of Animal and Plant Sciences,* 21(2 Suppl.):317–332. Available at: https://www.thejaps.org.pk/docs/21-2-suppl/JAPS-21(Final).pdf

Runciman, D. (2018) *How Democracy Ends*. Profile Books, London.

Younger, J. S. (1992) *Interim Report of the S2 (MSc) Programme in Highways and Transportation at ITB*. Institute of Technology Bandung, Bandung, Indonesia.

Younger, J. S. (2012) *Aceh, Eight Years on from Disaster*. Forbes, Indonesia.

Younger, J. S., Booth, D. J. and Kurniawan, K. (2012) Sustainable development: East Bali Poverty Project. *Proceedings of the Institution of Civil Engineers: Municipal Engineering*, 165:43–58.

Younger, J. S., Booth, D. J., Parry, D. E. and Kurniawan, K. (2017) Sustainable humanitarian development: East Bali Poverty Project. *Proceedings of the Institution of Civil Engineers: Municipal Engineering*, 170:67–77.

Younger, J. S., Sumiarso, L., Murray, M. I., Parr, D. E. and Prisandani, U. Y. (2022) Energy resources of Indonesia and their application. In *2nd Dundee Energy Forum: Evolving Global Energy and Resources Landscape*. University of Dundee, Scotland.

The Restless Earth

4

David Parry, Scott Younger, and Jasper Cook

4.1 INTRODUCTION

Sustainable development requires a positive interaction between the Earth and its population. An appreciation of the Earth's composition and history is important for understanding and fostering this interaction.

The Earth is one of a group of planets within our solar system whose position in orbit is governed by gravitational force associated with Sun. Earth is a fraction of the size of the Sun, a little less than 1%. The Sun has a mass 330,000 times greater than the Earth which, in turn, has an equator of a little over 40,000 km. The Sun can be considered a huge fireball, which the approximately round Earth orbits. The heat from the Sun is vital to our very existence.

The origins of our solar system lie within the universal "Big Bang," which is thought to have occurred about 13.8 billion years ago. The Earth itself formed, from the gradual accumulation of "Bing Bang" debris around the Sun, had taken on its current rocky structure by around 4.6 billion years ago. Since then, the Earth has cooled and accumulated and modified a gaseous atmosphere, able to support life as we know it.

Although life in terms of current flora and fauna (including humans) has an immediate relationship with, and dependence on, the nature of upper layers of the Earth and its atmosphere, it is important to appreciate the overall dynamic structure on which they depend.

DOI: 10.1201/9781003566274-4

4.2 THE RESTLESS EARTH

4.2.1 The Solar System

The solar system is only one of many within the billions of stars in the greater beyond, billions of miles into an apparent great infinity. It is interesting to draw a parallel with the more mundane, but more practically germane, question of sustainable development, which we can only visualize in a much smaller scale.

Although, based on the observation of other solar systems, it is anticipated that our Sun will gradually decay over the next tens of billions of years. However, it is assumed for the foreseeable future that the Earth will continue to spin in space, rotate around the Sun, and be influenced by the same natural events, that are with us today. It has to be borne in mind that the surface of the Earth today differs from that of a few million years ago and significantly compared with 1 billion years or more. Also, while changes are monitored, they have the potential to cause global devastation as evidenced by the several episodes of fauna and flora extinction within Geologic Time. What form of life will develop in the remaining Sun years? Will humans last that long and how will they evolve? Will they have gone to other planets as their curiosity continues to explore beyond Earth?

However, we must try to resolve for the foreseeable future, within two or three human life cycles, a measure of sustainability, which will then hopefully be taken care of, or upgraded by the generations which follow. There will have been an inevitable change in conditions, as we develop, mature as a species and as the population is altered. Perhaps the pace has quickened?

4.2.2 Earth Structure

The fundamental structure of planet Earth comprises three principal layers within an enveloping atmosphere (see Figure 4.1).

The Earth's rocky crust is the outermost layer and in direct contact with human activity. It varies in thickness, with oceanic crust being 6–11 km thicker and continental crust being thicker at around 30 km. Only in the past few years that have we learned to fully appreciate the crust, the vital importance it plays in everyday life and the necessary associated steps that we need to take to ensure a sustainable future.

The mantle underlying the crust extends to a depth of about 2,900 km and is composed of solid, primarily basic to ultra-basic. Although it is solid, the mantle material behaves as a very slow-moving viscous or plastic material.

The Earth's core is divided into two parts: the outer core and the inner core. The outer core, beneath the mantle, extends to a depth of about 3,500 km below the surface and is primarily composed of iron and nickel kept in a molten state by high temperatures and pressures. The gyratory motion of the molten iron in the outer core is considered to be responsible for generating the Earth's magnetic field. The inner core is situated at the

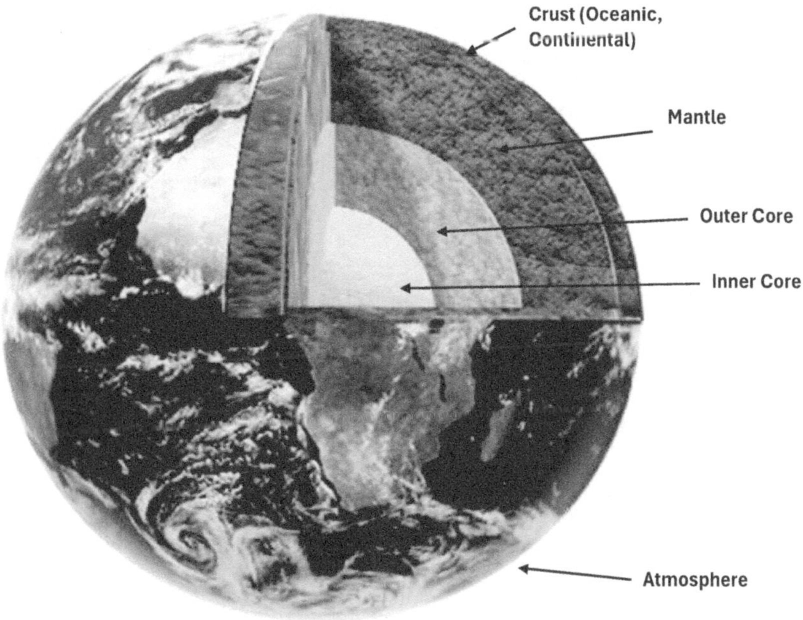

FIGURE 4.1 Major structural elements of the Earth.

very centre of the Earth below the outer core. It is also composed of solid iron and nickel but kept in a sold sate solid due to tremendous pressure.

4.2.3 Tectonic Plates and Hazards

The crust is made up of large tectonic plates in slow but constant motion over the underlying mantle. There are seven or eight major plates and a further five to six minor plates or platelets separated by a variety of plate boundary types as shown in Figure 4.2 and Table 4.1. The plates themselves are either continental or oceanic in nature. Pressure and stresses at these plate boundaries give rise to different tectonic adjustments depending on the nature of the boundaries. These tectonic hazards must necessarily be a factor in strategies for sustainable development.

Principal hazards impacting development that are associated with tectonic activity are as follows:

- Earthquakes
- Tsunamis
- Volcanic eruptions
- Lava sheet flows

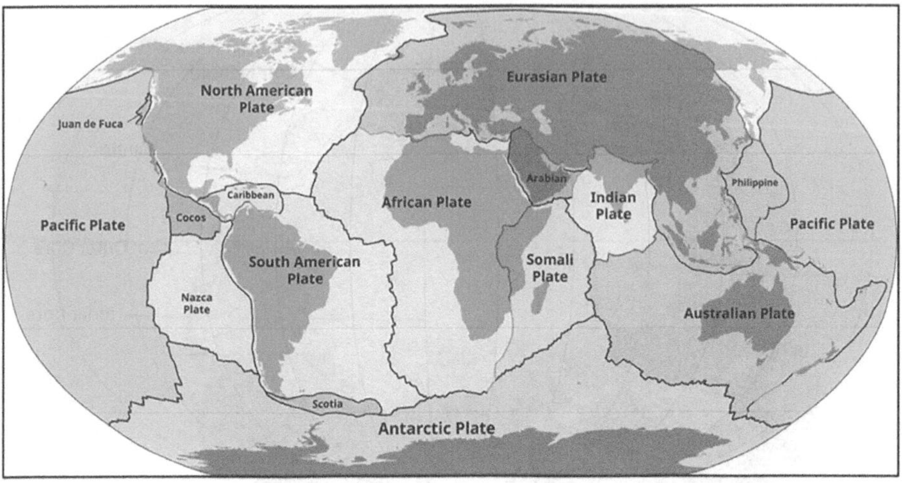

FIGURE 4.2 Major tectonic plates (Wikipedia, 2025).

TABLE 4.1 Tectonic plate boundary types

BOUNDARY TYPES	*PHYSICAL MODEL*	*PRIMARY GEOLOGICAL CONSEQUENCES*
Convergent	A. Oceanic crust subducted below a continental plate and melted at depth.	Explosive volcanic ash, gas and debris flows (Lahar). Limited viscous andesitic lava flows. Deep earthquakes.
	B. Continental crust impacting a continental crust.	Thrusting up of mountain belts and formation of metamorphic rocks at depth.
Constructive	Emergence of basic mantle material to act as a "conveyor belt" to split and move apart existing plates. Hotspots: igneous activity associated with isolated mantle activity.	Production of basaltic lava flows with some associated extrusive ash flows.
Transform	Periodic and sudden release of strain energy. No igneous activity.	Destructive shallow-wave earthquakes.

TABLE 4.2 Tectonic plate boundary types and hazards

OCCURRENCE	*TECTONIC ACTIVITY*	*CONSEQUENCE*
Icelandic igneous eruptions, 2024	Lava flow at **constructive plate** boundary (Mid-Atlantic Ridge).	Impact on transport and human habitation infrastructure.
Turkish–Syrian border, 2024	Two shallow earthquakes (Richter 7 and 7.7) related to faulting associated with **transform plate** boundary.	Widespread destruction of infrastructure and significant loss of life.
Sumatra, Thailand, Sri Lanka, 2004	Large tsunami resulting from a Richter 9.0 earthquake within the subduction zone of a **convergent (oceanic–continent)** plate boundary.	Widespread destruction of infrastructure and loss of around 286,000 lives.
Kashmir, 2005	A Richter 7.6 earthquake associated with the convergent **(continent–continent)** plate boundary.	Over 70,000 lives lost and over 35,000 buildings destroyed.

Table 4.2 summarizes the typical hazards and their impacts associated with types of plate boundary.

4.2.4 Earthquakes

When stresses imposed in the Earth's crust exceed the available strength of rock, or combinations of rock, they break, or move along pre-existing lines of weakness, and an earthquake results from the sudden release of strain energy. Earthquake seismic waves radiate outwards from the point failure or movement that is termed the focus, which may be many kilometres deep. The point at the surface directly above the focus is called the earthquake epicentre.

The magnitude of earthquakes is commonly measured by the logarithmic Richter scale (Richter, 1935). While the Richter scale continues to be well known by the general public, seismologists mainly use other more precise measurements, especially for extremely large earthquakes that more accurately measures the total energy released by an earthquake. However, most non-specialist groups still refer to the Richter scale for ease of understanding (Table 4.3). The Richter magnitudes are defined on a logarithmic scale; for example, a Richter 6 earthquake, which is moderately strong, is 10 times the power of an earthquake valued as Richter 5.

There are around significant 100 earthquakes each year of a size (>R6) that potentially could cause significant damage and thousands of minor earthquakes (<R4) that may go largely unnoticed. Major earthquakes strike largely without obvious warning, and as many of the Earth's earthquake zones coincide with areas of high population density, the results can be catastrophic, with high loss of human life and major economic cost.

TABLE 4.3 The Richter scale

SCALE	*SEVERITY*	*IMPACT*
1–1.9	Micro	Instrument recording only, not felt by humans. Estimated 1.5> million occurrences annually worldwide.
2–2.9	Minor	Often felt but rarely cause any damage. Around 1 million occurrences each year.
3–3.9	Light	Often felt but rarely causes significant damage. Around 100,000 of these quakes occur annually.
4–4.9	Moderate	Noticeable shaking, possible minor damage but major damage is unlikely. About 10,000 occurrences globally each year.
5–5.9	Strong	Potential to cause significant damage to buildings and other structures. Roughly 1,000 occurrences annually.
6–6.9	Very strong	A lot of damage and potential for significant loss of life in populated areas. About 100 occurrences each year.
7–7.9	Severe	Potential for causing serious structural damage and major loss of life depending on location. They happen around 10–20 times per year globally.
8–9	Violent	Major structural damage and potential for catastrophic loss of life, either directly or indirectly.
10	Extreme	No R10 earthquake in recorded history.

Most major earthquake zones are associated with plate boundaries but there are variable driving forces for earthquake activity remote from plate boundaries. These may result from the following:

- Reactivation of old geological fault lines
- Crustal uplift/depression as a result of changes in ice sheets
- Collapse of old mines
- Military activity (nuclear testing)

The damage and loss of life due to earthquake activity are not only a function of earthquake power, but also dependent on location with respect to population density and, most importantly, on the resilience measures in force in terms of infrastructure design and construction. Failure in enforcing regulations is considered to have been a key factor in the disastrous consequences of the Turkish–Syrian border earthquakes of 2023.

4.2.5 Tsunamis

A tsunami is a wave or series of waves in a water body caused by significant disturbance or displacement of a large volume of water by a large event. Most large tsunamis are caused by earthquakes at destructive plate boundaries. In deep ocean waters, tsunamis have small wave heights and a very long wavelength and generally pass unnoticed at sea. Their impact potential grows significantly when they approach shallow coastal waters, as the waves can grow in height by several metres as the wavelength shortens.

In addition to earthquakes, tsunamis may be caused by very large land slips, volcanic activity, glacier calving, or more rarely, in the geological timescale, meteorites/meteors.

Seiches are similar to tsunamis but take the form of standing waves in an enclosed or partially enclosed body of water. Seiches and seiche-related phenomena have been observed on lakes, reservoirs, swimming pools, bays, harbours, caves and enclosed seas. The key requirement for formation of a seiche is that the body of water be at least partially bounded, allowing the formation of the standing or resonating waves.

Seiches are not so large as tsunamis but can still be disruptive. For instance, Lake Baikul in Siberia, the largest freshwater lakes in the world, has experience of seiches as a result of strong earthquakes under the lake.

Both tsunamis and seiches can also be derived from volcanic eruptions. The largest historic tsunami was located below the Indian Ocean resulting from largest earthquake in modern times, assessed at Richter 9.0, off the coast of Sumatra, Indonesia, on the 26th December 2004 close to the island of Nias. The earthquake within the subduction zone of a destructive plate boundary was at fairly shallow depth and generated a huge tsunami wave that reached up to 20 m height at the coast. It affected much of the northernmost province of Aceh, destroying most of Banda Aceh before continuing onto Phuket in Thailand. It travelled to damage Galle in Sri Lanka and right across the Indian Ocean to make its presence felt on the East African coastline.

The devastation and loss of life was immense (some 286,000 people). Many world and national aid agencies and caring individuals rallied to assist the Government of Indonesia to rehabilitate and reconstruct Banda Aceh city and many areas of the province of Aceh.

4.2.6 Igneous Extrusion Events

Igneous rocks are largely derived from melts of existing rocks and magma either in the mantle or crust. Solidification into rock occurs either below the surface as intrusive rocks (such as granite) or on the surface as extrusive rocks (such as basalt). It is the process of extrusion of igneous rock resulting in volcanoes or lava flows that is of great significance in the context of sustainable development. On the positive side, igneous intrusions and extrusions have provided very significant development resources either as construction materials, geothermal energy, or in the generation of agricultural soil profiles. However, on the negative side, impacts from volcanic eruptions and lava flows can cause significant damage, loss of life and socio-economic set-back.

Igneous rock extrusion occurs in a number of geological settings, principally associated with plate boundaries or crust–mantle hotspot, mainly by explosive volcanic eruption or by lava flow from vents or rifts in the crust.

Volcanic eruptions are an awesome display of Earth's power that can cause disastrous loss of life and property, especially in densely populated regions of the world. One feared phenomenon accompanying some explosive eruptions is lahar, or pyroclastic flow, a fluidized mixture of hot gas, rock fragments and incandescent particles that sweep down a volcano's flanks, incinerating everything in its path. The most explosive volcanic events tend to involve the more viscous acidic (andesitic) magmas rather than the more easily flowing basic (basaltic) lava magmas. The former are characteristic of destructive plate boundaries and latter with constructive boundaries.

Perhaps the best-known ash eruption that took place was that of the stratovolcano Mt. Vesuvius in CE 70 which still lures archaeologists to the overwhelmed sites of the towns of Pompeii and Herculaneum. There had been warnings for the previous 8 years, but nothing to alert the people of the size of the eruption when it came. Pliny the younger was able to record the event in his diaries; he was staying at a nearby island but sufficiently remote as to be unaffected.

Indonesia has 141 active volcanoes, which is the most of any country in the world. Most of these also lie close to the edge of the Australian–Indian tectonic plate and the others are affected by the Pacific Ocean plate The well-documented Krakatoa eruption in the Sunda Strait between Java and Sumatra occurred in 1883, the fifth largest, only the eruptions of Mt. St. Helens in 1980 and Tambora being larger. Krakatoa remains very active, and the latest eruption took place at the tail end of 2018 when a shoulder broke off into the sea and raised a modest tsunami which killed some end-of-year holidaymakers across the Sunda Strait in Java. Eruptions have occurred in recent times at Mt. Merapi, Central Java, and also Mt. Rinjani, Lombok and Mt. Agung, Bali, with parallel earthquakes.

Mt. Agung affected the East Bali Poverty Project area (see Chapter 3) with over 100 homes destroyed. It produced the largest eruption of the 20th century and led to the deaths of 30,000 people and many of the community displaced to Lampung in S. Sumatra. Mt. Merapi in Central Java is arguably the most active volcano spewing ash every few years and refreshing the soil. The government in recent times has advised strongly residents to remain at least 8 km away.

Quite clearly, most of islands of Indonesia have a landscape formed by, and is at risk from, these large natural events such as earthquakes and volcanic eruptions, which in turn are related to the movement of two of the tectonic plates. These events are going to inevitably put back development, although major events, such as Aceh, do attract large emergency funds from overseas, otherwise costs are taken from government budget.

Flows of the less viscous basaltic lava can prove equally as destructive as volcanoes and cover much wider areas. Major lava flows covering thousands of square kilometres have in the geological past had enormous impacts on Earth history, including contributing to partial life extinction events through their effect on the atmosphere as well as direct physical impact (see Box 4.1).

BOX 4.1 THE DECCAN TRAPS

The 65 million-years-old Deccan Traps flow basalt is one of the largest volcanic features on Earth, cropping out over 500,000 km^2 of the west-central Indian subcontinent. The lava flow complex is predominantly composed of multiple lava flows of basalt, varying from more than 2,000 m in the west to over 1,000 m in the east and to less than 100 m in some southeastern regions. Some scientists consider that the gases released during the formation of the Deccan Traps played a supporting role in the Cretaceous–Tertiary or K–T extinction ("dinosaur" extinction) primarily caused by the Chicxulub meteor impact event.

A recent lava-flow event took place in Iceland, in 2023, where there was a 3 km long crack in the surface of a volcano with hot molten lava moving up from the core, and flowing slowly down towards the beach and sea, fortunately not in the direction of the town of Grundavik. Iceland is probably the single most tectonically active location in the world, sitting astride two tectonic plates (the Mid-Atlantic Ridge constructive margin).

Another recent lava flow event occurred in 2022 on the island of Mauna Loa (Hawaii). This location (over crustal hotspot) had been quiet for 40 years until erupting with streams of hot, molten lava travelling down the slopes.

4.2.7 Polar Reversal

A polar reversal occurs when Earth's magnetic north and south poles swap locations and is believed to be linked to changes in the magnetic behaviour within the iron-rich liquid outer core (Beggan, 2025). Reversals are common in Earth's geologic history and paleomagnetic studies indicate that they have occurred hundreds of times in the past 160 million years and have happened on average about every 300,000 years with the last one occurring around 780,000 years ago. While no one knows exactly when the next reversal will occur, some scientists believe we are long overdue for another reversal and that current "polar wandering" may be a precursor.

A reversal may be considered as being a rapid process in Geologic Time, but studies of reversals captured in volcanic eruption sequences indicate the process takes on the order of 5–10,000 years. Our magnetic field provides some protection from the impact of solar and cosmic radiation; as the field weakens possibly by up to 90% during a pole reversal, there is the possibility that more increased radiation could reach Earth.

The last reversal occurred long before recorded human history, and very little can be identified regarding effect of a reversal on current environment. The consequences of a pole flip unlikely lead to extinction events, but there is a potentially high risk to the electronic-based developed lifestyle as we know it (Hardin, 2025). Potentially serious sustainability consequences that should at least be identified and planned for can be summarized as follows:

1. Power grid disruptions: a weakened magnetic field can cause stronger geomagnetically induced currents in power grids resulting in long-term and widespread blackouts across national power systems.
2. Telecommunications interruption: disruption of satellite communications, GPS systems and radio signals.
3. Electronic failures: malfunction of electronic equipment, both on Earth and in orbit.
4. Increased radiation: greater human and other animal life exposure to solar radiation and cosmic rays could cause increased risk of radiation-related health issues (including cancer).
5. Environmental disturbances: a weakened magnetosphere may potentially lead to significant changes in ozone levels, weather patterns and general climate changes.

4.3 THE EARTH'S ATMOSPHERE, WEATHER AND CLIMATE

4.3.1 Introduction

Before we can understand climate change or even what climate and the atmosphere mean, we need to travel back to the beginning of the Earth's history and appreciate some fundamentals of celestial mechanics including the meaning of such terms as "gravity." This is the force that attracts a body towards the centre of the Earth or towards any other physical body having mass. Most school children know the largely apocryphal "falling apple story" on how gravity was identified by Sir Issac Newton. What is true is that he did define his universal law of gravitation (1687) as the mutual attraction of massive bodies and the force that keeps the planets orbiting around the Sun without being drawn into it, thus reinforcing Copernicus' theory that the Sun was at the centre of our solar system.

Cosmologists generally use the "big bang" theory to explain the origin of the universe and our Solar System within it (see Section 4.2). The Earth orbits around the Sun in an anticlockwise direction (east to west) every 365 days; however, the axis of rotation of the Earth is not perpendicular to the plane of its orbit around the Sun but is tilted by an angle of 23°. This tilt was caused by a planet the size of Mars, impacting the Earth some 4.5 billion years ago sending pieces of both celestial bodies flying into space. The impact would have left a molten Earth that began to cool rapidly, while the debris from the collision settled into orbit around the Earth under the influence of gravity and eventually coalesced to form the Moon.

Axial tilt means that as the Earth revolves around the Sun, the north pole is tilted towards the Sun on June 21st (summer solstice) and away from the Sun on December 21st (winter solstice). The Sun crosses over the celestial equator on March 21st (vernal equinox) and September 21st (autumnal equinox). It is this variation in axial tilt that creates our seasons combined with other orbital wobbles such as eccentricity and precession that cause cyclical variations in the distribution of solar radiation at the Earth's surface and this orbital forcing strongly influences the Earth's climatic patterns. These Milankovic cycles, which occur over thousands of years, are named after the Serbian geophysicist and astronomer, Milutin Milankovic who identified a 100,000-year cycle for the orbital elliptic, a 41,000-year cycle for axial tilt and a 27,500-year cycle for axial precession. Without this orbital tilt, there would be extremes of heat and cold at the equator and poles, respectively. On top of these long-term variations in the amount of sunlight reaching the Earth, there are other short-term cycles that affect temperature and rainfall. The most significant, the El Niño–Southern Oscillation, involves circulation changes in the tropical Pacific Ocean on a time frame of 2–7 years that strongly influence rainfall in North America, Africa, Australia and Southeast Asia. The North Atlantic Oscillation and the Indian Ocean Dipole also produce strong regional effects. Both of these interact with the El Niño–Southern Oscillation.

The interconnections between these cycles used to make it hard to show that human-caused climate change was statistically significant and not just another lurch of natural variability. But anthropogenic climate change has demonstrably gone beyond natural variability within human timescale in terms of weather and seasonal temperatures (Santer et al., 2013). The U.S. National Climate Assessment (USGCRP, 2017) concluded that there is "no convincing evidence for natural cycles in the observational record that could explain the observed changes in climate." It should be borne in mind, however, that large-scale climate changes have frequently occurred in geological time. Significant Ice-Age climatic variations occurred within the last 100,00 years had a significant impact on human development. These and earlier events within the last billion years occurred without any anthropological influence.

4.3.2 The Atmosphere

There is often confusion among laypeople about what constitutes the atmosphere and the differences between weather and climate, with the latter two terms seemingly interchangeable. Any meaningful appreciation of the climate debate must begin with an understanding of these terms. Weather and climate are not the same thing. Weather reflects short-term (daily, weekly, monthly) atmospheric conditions such as the rain, wind, sunshine and barometric pressure. Climate refers to the long-term regional or global average of these conditions and may be considered over decades and even longer into Geologic Time.

Weather is very much part of our everyday lives, and the likely local weather forecast for a day or so ahead is provided by television, radio and now, the internet. The permutations of weather and their effect on our daily routines are endless. The forecast has a major impact on what we immediately do and how we do it. Be it planning agricultural, commercial fishing, or leisure events. Climate forecasting is equally if not more important, impacting as it will on things such as agricultural strategy or the performance of major Civil Engineering structure over a design lifetime of 100 years or more.

Let us consider the atmosphere since that is where weather and climate are born. The atmosphere is the layer of gases retained by Earth's gravity that surrounds the planet from the surface to 100,000 km or thereabouts into space although 85% of the atmosphere is contained in the first 10,000 km. The atmosphere protects life on Earth by creating pressure allowing for liquid water to exist on the Earth's surface, absorbing ultraviolet solar radiation, warming the surface through heat retention (greenhouse effect) and reducing temperature extremes between day and night.

The gaseous composition of the atmosphere is about 78% nitrogen, 21% oxygen, 0.9% argon and 0.1% other gases including small amounts of carbon dioxide (0.04%), methane and water vapour. The latter can occupy 2% or more of the atmosphere under warm and humid conditions when the percentage of the major gases, nitrogen and oxygen, will drop.

Water vapour accounts for most of the atmospheric greenhouse effect with carbon dioxide the next most important, igneous eruptions, the burning of fossil fuels and deforestation and the one that human activity has the most impact on carbon dioxide enters the atmosphere through the decay of carbon products. It is removed from the atmosphere by the oceans and photosynthesis, a process used by plants and other organisms to convert light energy into chemical energy. A basic understanding of the mechanism is a prerequisite in the climate debate.

4.3.3 Photosynthesis

Photosynthesis is a system of biological processes by which organisms, such as most plants, algae and some bacteria, convert light energy, typically from sunlight, into the chemical energy. Some of the chemical energy is stored in carbohydrate molecules such as sugars and starches, which are synthesized from carbon dioxide and water. The process always begins when energy from light is absorbed by proteins called reaction centres that contain green chlorophyll and other coloured pigments called chloroplasts. These are most abundant in leaf cells. Some light energy is used to strip electrons from suitable substances such as water, producing oxygen gas. Photosynthesis is largely responsible for producing and maintaining the oxygen content of the Earth's atmosphere and supplies most of the energy necessary for life on Earth. Today, the average rate of energy capture by photosynthesis globally is 130 terawatts which is about 8 times the current power consumption of human civilization. Photosynthetic organisms convert around 100–115 billion tons of carbon into biomass per year.

Photosynthesis is vital for climate processes as it captures carbon dioxide from the air and then binds carbon in plants, in soils and in harvested products. Nonetheless, human activities, especially the burning of fossil fuels, have been responsible for increasing the level of carbon dioxide in the atmosphere by over 42% since 1750 when the concentration was 280 ppm. Then the population was less than 1 million. It increased to 315 ppm, a rise of 11%, in 1960 when the population was 3 million but rose to 425 ppm at the present time when the population has exceeded 8 million. This is an increase considerably greater that would have occurred due to natural forcing mechanisms which are discussed below.

4.4 CLIMATE AND CLIMATE CHANGE

4.4.1 Key Principles

The Intergovernmental Panel on Climate Change (IPCC, 2014) the premier international agency on climate science, defines Earth's climate as a complex, interactive system consisting of the atmosphere, land surface, snow and ice, oceans and other bodies of water, and living things. Climate is usually described in terms of the mean and variability of air temperature, precipitation and wind over a period of time ranging from months to millions of years. It is the long-term aggregation of weather for a given area that will vary in size from a few hundred square kilometres, to many thousands of square kilometres and, in some instances such as the Arctic Tundra, to millions of square kilometres. For the study of climate and how it has changed over time and is still changing, the scientific community requires a minimum of 30-years of meteorological observations from the world's weather stations to provide statistical significance to data analysis. It is the study of climate or climatology that is at the heart of the global warming debate and the climate predictions for the next century that the climate models of the developed nations produce (World Bank, 2024).

The climate system evolves in time under the influence of its own internal dynamics and as a result of external factors that cause or "force" changes. Such external forcing factors may be natural phenomena such as volcanic eruptions and solar variations, as well as human-induced changes in the gaseous composition of the atmosphere.

The climate and weather systems are powered by solar radiation which determines insolation, or the amount of solar radiation received on a given surface in a given time period. The balance of solar radiation can be changed in three fundamental ways:

1. By changing the incoming solar radiation, for example, by changes in the Earth's orbit (see Milankovic cycles) or in the Sun itself such as sunspot activity.
2. By changing the Earth's "albedo," or fraction of solar radiation that is reflected back into space, for example, from changes in cloud cover, snow and ice, atmospheric particles or vegetation.
3. By altering the infrared radiation from Earth back towards space, for example, by changing concentrations of the gases that trap heat in the atmosphere, known as the greenhouse effect and without which, life would not be possible since the planet's average air temperature would be about –200°C. The Earth's main greenhouse gases (GHGs) are water vapour (H_2O), carbon dioxide (CO_2), nitrous oxide (N_2O), methane (CH_4), ozone (O_3) and chlorofluorocarbons.

Climate responds directly and indirectly to such changes through a variety of feedback mechanisms that can either amplify ("positive feedback") or diminish ("negative

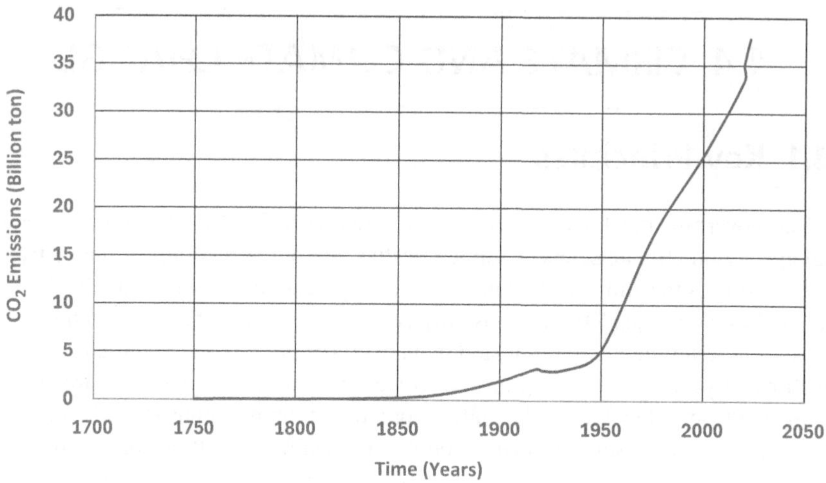

FIGURE 4.3 Historic CO_2 emissions (Ritchie, 2025).

feedback") the effects of a change in climate forcing. Detecting, understanding and accurately quantifying these feedbacks are a key focus of current climatological research.

Human activities, primarily the burning of fossil fuels and clearing of forests, have greatly increased the amount of carbon dioxide in the atmosphere from around 280 ppm in the middle of the 18th century to 425 ppm today.

The link between CO_2 emissions and industrialization is demonstrated in Figure 4.3, which shows dramatically the change from near-zero emissions in 1750 (estimated at 9 million tons) to those today (38 billion tons), the very sharp rise in emissions in the last 100 years, particularly the past 60+, being only interrupted for a short time by the Great Depression (1930–1935) and the COVID pandemic (2020–2021).

This increases the amount of water vapour in the atmosphere, intensifying the natural greenhouse effect causing enhanced global warming. As the Earth warms, snow and ice begin to melt revealing darker land and water surfaces beneath the snow and ice which absorb more of the Sun's heat, causing more warming which cause more melting and so on, in a self-reinforcing cycle. This ice-albedo feedback amplifies the initial warming caused by rising levels of GHGs.

4.4.2 The Misconception of a World Climate

Climatologists base their research on global warming largely on a mean world air temperature. This has led, among laypersons in particular, to the concept of an average world climate. Yet common sense tells us that at any one time, meteorological conditions vary greatly throughout the world from extremes of heat to extremes of cold and everything in between. There are sound scientific reasons for using the average world air temperature as a benchmark, but it is important to put this into context of the non-average climate conditions in which the human race lives.

4.4.3 The Practical Advantages of Recognizing Regional Climates

Climate is perhaps it is best summed up as the long-term (>30-year) aggregate of meteorological or weather phenomenon such as rainfall, temperature, sunshine (insolation), wind and humidity that affect every square metre of the planet's surface. These data are aggregated and compared to detect meaningful and statistically significant differences over a large area (>100,000 km^2) that can be designated as regional climates. These regional climates that occur over the planet's land masses and oceans, play a huge part in our daily lives and in shaping the Earth's landscapes.

The concept of climatic zones on the Earth was first postulated by the Ancient Greeks with their latitudinal division of the Earth into frigid, temperate and torrid zones north and south of the equator. Many other systems of climate classification have been formulated since then but all of them may be broadly differentiated as using either empirical or genetic methods. Empirical methods make use of observed environmental data such as temperature, humidity and precipitation. In contrast, a genetic method classifies climate on the basis of its causal elements that give rise to the spatial and temporal patterns of climate data. Such elements include, among others, air mass circulation systems, jet streams, solar radiation and topographic effects.

Today, the most widely used climate classification system is the empirical one devised by the German botanist and climatologist Wladimir Köppen, first published in 1900, revised in 1918 and revised again before his death in 1940 (Köppen, 1936). It was later updated and revised by Rudolf Geiger in 1961 as shown in Table 4.4, to be followed in 2006 by an updated digital version of the map based on recent data sets from the Climate Research Unit at the University of East Anglia and the Global Precipitation

TABLE 4.4 Köppen–Geiger climate classification scheme symbols

FIRST	*SECOND*	*THIRD*
A (Tropical/Equatorial)	f (Rainforest) m (Monsoon) w (Savanna, Dry Winter) S (Savanna, Dry Summer)	
B (Arid)	W (Desert) S (Semi-Arid Steppe)	h (Hot) k (Cold)
C (Temperate)	w (Dry Winter) f (No Dry Season) S (Dry Summer)	a (Hot Summer) b (Warm Summer) c (Cold Summer)
D (Continental)	w (Dry Winter) f (No Dry Season) s (Dry Summer)	a (Hot Summer) b (Warm Summer) c (Cold Summer) d (Very Cold Winter)
E (Polar)	T (Tundra) F (Eternal Frost – Ice Cap)	

Climatology Centre at the German Weather Service (Kottek et al., 2006). It is believed that the new digital Köppen–Geiger world map on climate classification is valid for the second half of the 20th century. The Köppen–Geiger classification identifies 32 regional climates throughout the world as shown in Table 4.5 (Beck et al., 2018).

Increasing temperatures, rising sea levels, changes in wind and precipitation patterns that are related to human-driven (or "anthropogenic") climate change will affect these regional climates in different ways and to differing degrees. In some regions, increasing air temperatures will be the main parameter of change, while in others increasing or decreasing levels of rainfall will occur, possibly accompanied by changes in rainfall intensity, frequency and timing. Before considering the implications of climate change for the purposes of mitigation or adaption, the first task is to define precisely the location under study and the regional climate in which the location sits. Two examples at opposite ends of the climate spectrum will serve to illustrate the point.

Nur-Sultan (formerly Astana), the capital of Kazakhstan, is the second coldest capital city on Earth after Ulaanbaatar in Mongolia. The city has an average annual temperature of 3.5°C and winter temperatures as low as –35°C. The Esil River that flows through the city centre remains frozen from mid-November to the beginning of April and outdoor work is effectively stopped for the December–March period due to extreme

TABLE 4.5 Köppen–Geiger regional climates

1. Equatorial climate (Af)	16. Humid subtropical/subtropical oceanic highland climate (Cwb)
2. Monsoon climate (Am)	17. Oceanic subpolar climate (Cwc)
3. Tropical savanna climate with dry winter characteristics (Aw)	18. Warm continental/humid continental climate (Dfa)
4. Tropical savanna with dry summer characteristics (As)	20. Temperate continental/humid continental climate (Dfb)
5. Warm desert climate (BWh)	21. Cool continental/subarctic climate (Dfc)
6. Cold desert climate (BWk)	22. Subarctic/boreal climate (Dfd)
7. Warm semi-arid climate (BSh)	23. Warm continental/Mediterranean continental climate (Dsa)
8. Cold semi-arid climate (BSk)	24. Temperate continental/Mediterranean continental climate (Dsb)
9. Warm Mediterranean climate (Csa)	25. Cool continental climate (Dsc)
10. Temperate Mediterranean climate (Csb)	26. Cold continental climate (Dsd)
11. Mediterranean cold summer climate (Csc)	27. Warm/humid continental climate (Dwa)
12. Humid subtropical climate (Cfa)	28. Temperate/humid continental climate (Dwb)
13. Temperate oceanic climate (Cfb)	30. Cold continental/subarctic climate (Dwd)
14. Cool oceanic climate (Cfc)	31. Tundra climate (ET)
15. Dry-winter humid subtropical climate (Cwa)	32. Ice cap climate (EF)

cold and frozen ground. The climate of Nur-Sultan under the Köppen–Geiger system classifies as continental (D) with a dry (w) and a very cold winter (d), equivalent to regional climate 30 in Table 4.5.

An increase in the average surface air temperature by 3°C, say, for the second half of the 21st century, would bring the average temperature up to 6.5°C. This is still cold by some standards (most North European cities have average temperatures in the low- to mid-teens), and below the world average surface temperature of 15°C, but could improve living conditions and the economy of Nur-Sultan considerably by shortening the period when the ground is frozen, lengthening the summer growing season for crops and reducing the heating requirements for housing.

At the other end of the climate spectrum, such a temperature spike in Asmara, the capital of Eritrea, which has a tropical savannah climate with a dry winter (Aw) and a mean annual temperature of 17.1°C, would take the temperature spike to 20.1°C, akin to temperatures in today's monsoonal climates but without the attendant rainfall. Such changes would have huge ramifications for agriculture, the economy and the population, spurring crop changes and an increase in electrical power for residential cooling and accelerating water conservation measures. Clearly all predicted changes in the mean global temperature need to be set against the regional climate in which they will occur for effective mitigation and adaption measures to be implemented.

The science behind why human activities, especially the burning of fossil fuels, have had, and continues to have, such an impact on the average air temperature of the Earth is well understood. It is only the climate forcing factor or positive feedback over which humanity appears to have some control, hence the sense of urgency among environmentalists and belatedly among governments, to drastically curb GHG emissions to prevent the planet warming by more than 1.5°C before the end of the 21st century.

Unfortunately, this sense of urgency manifests itself only among certain governments, industries and commerce.

Since this circus has gone on for around 27-years with so little to show for it, is it not time to re-think the whole approach to climate mitigation through the UN? A route that, with a little thought, would have been deemed impossible in the first place given the difficulty of achieving any form of consensus among member states on any issue, let alone one as important and far-reaching as climate. This is not to demean the genuine efforts of thousands of dedicated government officers tasked with brokering climate deals, but rather to draw attention to the inadequacy and inappropriateness of this approach. Rather than striving for agreement between governments, agreement on climate issues within governments and, by association, within national populations, is likely to be more successful and have a positive knock-on effect globally.

4.4.4 Climate-Related Hazards

Climate-related hazards have historically mostly been linked with equatorial or subequatorial locations, for example the Caribbean and Southeast coast of the USA, South Asia, South China Sea locations, and in the Pacific. However, this restricted view is no longer valid as significant climate events are increasingly reported from North America, Europe

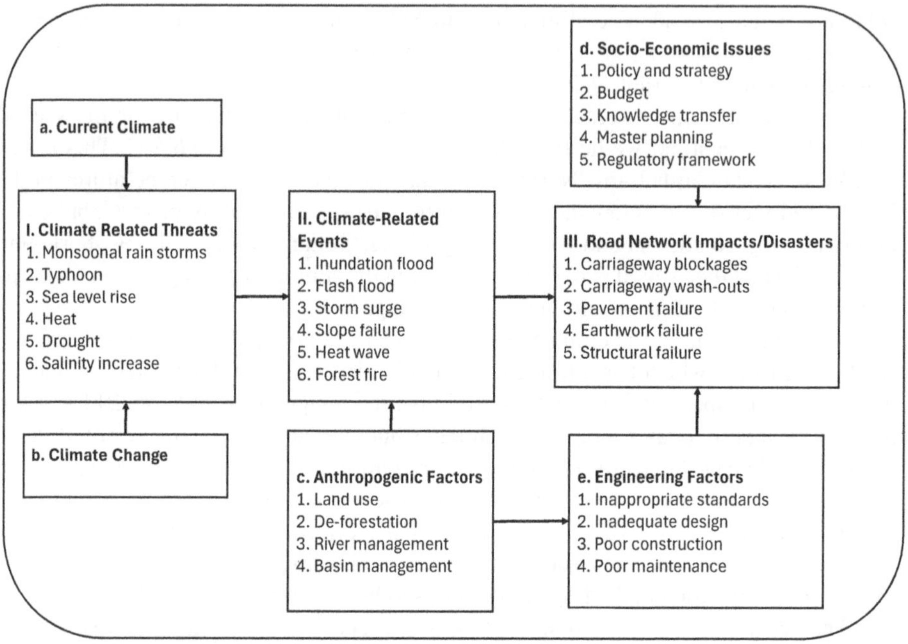

FIGURE 4.4 Climate impact framework related to transport infrastructure (Cook, 2022).

and Australasia. Climate-related hazards are now of world-wide concern. As an illustration of the inter-relationship of climate hazard factors, Figure 4.4 summarizes key linkages between climate hazards, impacts and consequences in the road transport sector.

4.4.5 Mitigation

There are two fundamental approaches to dealing with climate change. The first is mitigation, the path favoured by the UN and IPCC that actively, through good will, common sense and collective responsibility, but ultimately through legislation, seeks to reduce anthropogenic GHG emissions, especially of carbon dioxide, the GHG that is forcing temperature increases worldwide.

4.4.6 Adaption

The second approach is adaption whereby localities, legislatures and individuals accept the reality of global warming (and by implication, the failure of the mitigation approach to the problem) and seek ways and means to adapt their environment and way of life to the expected temperature and sea-level rises. For example, adaption uses civil engineering and bio-engineering techniques, among others, to strengthen existing seacoast defences and change land-use practices in order to lessen the impact of the threats.

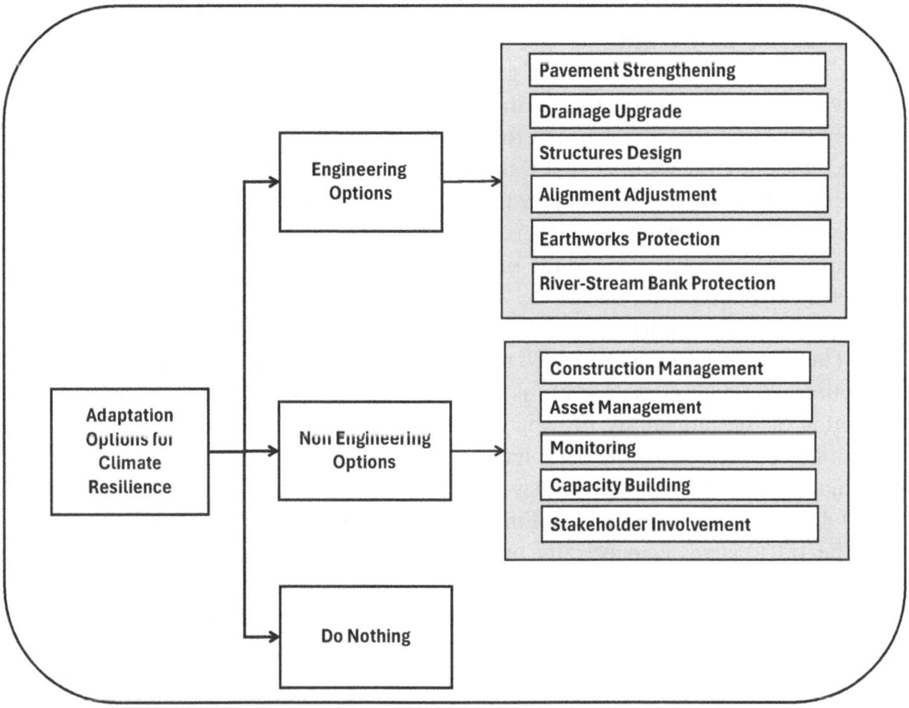

FIGURE 4.5 Common adaptation measures in the context of transport infrastructure (ADB, 2011).

Figure 4.5 presents an example of adaption factors commonly applied to transport infrastructure development (Cook, 2022).

In reality, both approaches are necessary, working in tandem to reduce GHG emissions and at the same time, plan and implement adaption measures.

4.4.7 Climate Change Paradox

The scientific case for global warming under the all-encompassing soubriquet, climate change, is irrefutable, and the clamour for governments, corporations and individuals to act on the science from the UN and the IPCC is both strident and persistent and yet, as the continuing failure over the past 30 years to make any serious reduction in GHG emissions demonstrates, the message is largely falling on deaf ears. Understanding this paradox will go a long way to opening the door for real progress to be made along the mitigation path before it is too late or, in popular climate science parlance, a tipping point is reached beyond which recovery will be impossible and runaway heating will occur.

We have a unique situation. Generally, the results and recommendations of scientists in any field are believed and acted upon by governments and civil society. One only has to think of medicine, nuclear science, astrophysics, genetics and biology. However, it appears that climate science is not entirely trusted by society at large, including governments. This has nothing to do with the fossil fuel industry funding advertising and persuading governments to look the other way, while they continue to make their profits from producing coal, oil and gas to power the world's economies. It is perhaps more visceral and even spiritual in that it goes back to before the intellectual renaissance of the 14th and 15th centuries, to a time when the human race was at the mercy of nature, especially the weather, and was powerless either to understand the phenomenon or do anything about it other than try to adapt and survive.

As scientific understanding progressed, the physics and chemistry creating weather phenomena became known, and the incredibly complex nature of weather and climate became clear. A quote from the IPCC 2021 Executive Summary report sums up the current thinking of the climate scientists admirably.

"The climate system is a coupled non-linear chaotic system and therefore long-term predictions of future climate states is not possible. Rather the focus must be upon the prediction of the probability distribution of the system's future possible states by the generation of ensembles of model solutions. Addressing adequately the statistical nature of climate is computationally intensive and requires the application of new methods of model diagnosis, but such statistical information is essential" (IPCC, 2021).

The IPCC states categorically that the coupled non-linear chaotic nature of the climate system precludes long-term predictions of the climate. Yet volumes of text are taken up in all the IPCC Reports with projecting likely climate scenarios 30, 50 and 100 years into the future. Is it any wonder that there is confusion among laypersons, policymakers and governments as to what the climate experts really mean? Add to this the public distrust of weather forecasts in general that go beyond 48 hours, largely because of the numerous failures exemplified, for example, in the British weather service's egregious failure to predict the "Great Storm" on 15 October 1987 (see Box 4.2).

BOX 4.2 THE GREAT STORM UK, 1987

Forecasters had predicted some days before that stormy weather was on the way, but models suggested the severe weather would pass to the south of England. It did not and the hurricane-force winds (>64 knots for extended periods) caused billions of pounds worth of damage to buildings, felled 15 million trees and seriously damaged transport and power infrastructure. In addition, 18 people were killed during the storm. Subsequent analysis of the storm declared it to be a one in 200-year event.

This failure and many others that have been recorded in all weather services do not nullify the excellent track record of most weather services in short-term forecasting, upon which we rely on to ensure we are suitably attired to carry out our daily activities. But it is human nature to remember the failures rather than the successes and thus add to the extensive folklore surrounding weather and climate, and to date mankind's ineffective and, in the eyes of some, misguided attempts to control the weather and by implication the climate.

4.4.8 Time for a Different Approach

This takes us to the very heart of the climate crisis and sustainable development, for unless we decouple the detailed modelling prediction dilemma from the unequivocable basic science proving well beyond any reasonable doubt that human activities are accelerating the heating of the atmosphere beyond that which would occur naturally, our future could be unsustainable.

To achieve this decoupling and focus on reducing our carbon footprint will require a paradigm shift in how we teach and report climate science with the emphasis on the physics and chemistry of the atmosphere and the biosphere and the hydrosphere with climate modelling reduced almost to a footnote for those interested in learning more. One only has to think of the highly complex modelling performed by physicists studying subatomic particles at the Large Hadron Collider in Switzerland. When a new particle like the pentaquark or the Higgs boson is discovered, detailed reports are written for the cognoscenti and articles published in the *Nature* journal for a wider audience, but the public are not bombarded with reports on the models, and the mathematics and physics behind the models that the IPCC publishes. Climate modelling should be studied and applied only by those with the necessary skills to solve or work around the fundamental flaw in climate modelling referred to by the IPCC (2014).

Further quotes from the Executive Summary of the Policy Makers Report (IPCC, 2014) demonstrate the more practical direction that climate modelling should take.

1. Improve the integrated hierarchy of global and regional climate models with a focus on the simulation of climate variability, regional climate changes and extreme events.
2. Link models of the physical climate and the biogeochemical system more effectively, and in turn improve coupling with descriptions of human activities.

At present, human influences generally are treated only through emission scenarios that provide external forcings to the climate system. In future more comprehensive models, human activities need to begin to interact with the dynamics of physical, chemical and biological subsystems through a diverse set of contributing activities, feedbacks and responses.

The focus on regional climates, likely changes in the same and frequency of extreme weather events occurring, represents a practical approach to implementing local adaptation strategies.

In parallel with climate modelling, climate scientists and the IPCC have been advocating international action to limit temperature rise to 1.5°C above pre-industrial levels since COP21 in Paris in 2015. Such has been the degree of failure to come anywhere close to achieving this target than some observers including, climate science luminaries like Bill McGuire, Professor Emeritus of geophysical and climate hazards at University College London (UCL) and author of *Hothouse Earth: An Inhabitant's Guide* (McGuire, 2022), are urging COP27 to acknowledge that this target is dead since there is no credible pathway to achieving it. To do so would require GHG emissions to fall by 45% in the next 7 years when they are actually on track to rise by 10%, compared with 2010 levels. McGuire (2022) goes onto argue that continuing to posit the viability of 1.5°C[1] is misleading and raises false hopes. As such, it is vital that COP27 squashes claims that the goal is still alive and acknowledges the fact that missing this critical target represents a colossal failure of the whole COP apparatus.

McGuire argues that in retrospect, the obsession of the IPCC and governments with a specific target, rather than fighting to stop every fraction of a degree in temperature rise, has been counterproductive for, as with all targets, they are always still reachable, until they are not. Our efforts at creating a sustainable planet for ourselves and future generations must begin by accepting and embracing this fact. Only when we do that, can we individually and collectively begin implementing the twin strategies of GHG mitigation and adaption to our warming world. Rather than striving for agreement between governments through the UN, which has proven to be a total failure, agreement on climate issues within national governments, local governments and, by association, within local populations, is likely to be more successful. This will have a positive knock-on effect globally, particularly since the population is expected to stabilize over the latter part of this century.

4.5 SOIL – THE LAYER OF LIFE

4.5.1 Introduction and Background

About 71% of the Earth's surface is water with the remaining 29% made up of the land. Currently about 8 billion people live on the land, existing with all the necessary infrastructure within the top 50 m or so of the surface. This layer generally consists of fresh to weathered soil-rock profiles overlain in most continental areas by a layer of agricultural soil. This frequently abused layer is responsible for life on Earth as we know it through the development of agriculture some 10,000 years ago. But what does this vital and seemingly magical layer, often referred to as dirt or Earth, consist of and why is it so abused by humankind?

There are numerous definitions of soil, largely depending on the scientific focus. The Oxford Dictionary defines it as the upper layer of Earth in which plants grow, typically consisting of a mixture of organic remains, clay and rock particles. A geological describes soil as the accumulation of loose weathered material that covers much of the land surface of the Earth to a depth ranging from a few millimetres to many metres (Whitten and Brooks, 1972). A newer definition has soil as the biologically active and porous medium that has developed in the upper-most layer of the Earth's crust. Soils vary greatly in their chemical and physical properties and provide the structural support to plants. Whatever the detailed definition, soil can be considered as serving as a reservoir of water and nutrients and a medium for the filtration and breakdown of injurious wastes. It also helps in the cycling of carbon and other elements through the global ecosystem.

Perhaps a further word should come from the Soil Survey Manual of the United States Department of Agriculture (USDA, 2017) that states, "The following general definition of soil may serve those who need one: Soil is the collection of natural bodies occupying portions of the Earth's surface that support plants and that have properties due to the integrated effect of climate and living matter, acting upon parent material, as conditioned by relief, over periods of time."

Soil is a mixture of organic matter, minerals, gases, liquids and organisms that together support life. Soil consists of a solid phase of minerals and organic material (the soil matrix), as well as a porous phase that holds gases (the soil atmosphere) and water (the soil solution). Thus, it is a three-state system of solids, liquids and gases. Soil is a product of several factors: the influence of climate, relief (elevation, orientation and slope of terrain), organisms and soils' parent materials (original minerals) interacting over time. It continually undergoes development by way of numerous physical, chemical and biological processes, which include weathering with associated erosion. Given its complexity and strong internal connectiveness, soil is regarded by soil ecologists as an ecosystem. The Earth's body of soil called the pedosphere, interfaces with the lithosphere, the hydrosphere and the biosphere. Much of the upper layers of Earth's soil are younger than the Pleistocene epoch and, apart from indurated "fossil" soils, none is older than the Cenozoic era.[2]

Soil science has two basic branches of study: edaphology and paedology. Edaphology studies the influence of soils on living things, and paedology focuses on the formation, description (morphology) and classification of soils in their natural environment. In engineering terms, soil is included in the broader concept of regolith, which also includes other loose material that lies above the bedrock, as can be found on the Moon and other celestial bodies. The innate complexity of the soil ecosystem means that there are tens of thousands of different types of soil on the Earth occupying a few square metres at the smallest hierarchical scale (the soil phase), up to thousands of square kilometres at the highest scale (the soil order). Despite this complexity, and to the everlasting benefit for humankind, plants' requirements from the soil for successful growth are not that challenging and the soils in the regions of the Earth where the majority of the population live, provide these requirements which can be summed up as follows: a warm humid climate for 6 or more months of the year and a metre or more of weathered slightly acid to slightly alkaline soil with organic material in the form of humus in the top 10cm.

If the great civilizations of the world had one common denominator, it was an ability to support a thriving agriculture to feed an ever-expanding population within a generally benign (humid and warmish) climate. The key to the sustainability of early civilizations was the presence of highly fertile renewable soils. The Sumerian (c.4500–1900 BCE) and Babylonian (1600–562 BCE) civilizations of the Tigris and Euphrates Valleys relied on the annual flooding of the two rivers to deposit another millimetre or so of new alluvium in their rich floodplains. Similarly, the ancient Egyptians (4th millennium BCE) were dependent on the floodwaters and silt of the Nile River for annual soil replenishment. The erosion of rocks and soils many hundreds of miles away in the watersheds of these three great river systems provided the source of the silt that helped these civilizations to flourish. But water is only one of a number of ways in which new soil material can be deposited. In China, vast tracts of soil many metres thick are derived from 4,000-year-old wind-blown loess deposits, one of the most fertile soil materials known. The north-western loess region is credited with being the birthplace of Chinese civilization.

In Java, the process of soil renewal was, and still is, related directly to the island's history of volcanism whereby over millennia the elements for new soil have been ejected violently in volcanic eruptions of ash or more gently as slow-moving and spreading lava flows. Under the prevailing moist and hot climate typical of Java for the past 3,000 years, these volcanic ash and lava deposits, rich in most of the minerals required by arable crops and natural vegetation, have been weathered to produce some of the world's most fertile and physically stable soils. The great Javanese Kingdoms of Saliendra (8th century CE) and Majapahit (c.1293–1527 CE) were built on such soils, and it is no coincidence that Borobudur (9th cenutry AD), one of the oldest and largest Hindu-Buddhist temples, is built within the shadow of Mount Merapi, one of the world's most active volcanoes. Typically, small eruptions occur every 2 or 3 years and larger ones every 10–15 years. Catastrophic eruptions resulting in great loss of life occurred in 1006, 1786, 1822, 1930 and 2010. The volcano erupted again in 2018, 2021 and 2023 causing the evacuation of thousands of villagers and their livestock. Despite the tragic loss of life during major eruptions, it is unlikely that local people will move away from the reach of the volcano for the same reason that their ancestors refused to move, namely the benefits in terms of soil fertility that the volcano brings.

The farmers of Merapi and those of other active volcanic areas have an empathetic relationship with their soil and the circumstances surrounding its evolution; such a relationship was generally the norm prior to the agricultural and industrial revolutions of 18th- and 19th-century Europe and the Americas when soil became a resource to be exploited without thought for its long-term health. The use of fallows and ploughed-in green manures, spawned by the agricultural revolution, was soon overtaken by monoculture, hedge-row destruction, mechanization, excessive chemical fertilization and the application of chemical herbicides and insecticides in pursuit of high yields and high profits. By the time the deleterious impacts such practices were having on the soil were seen and, in some measure, understood, for the southern plains of the USA in the 1930s, it was already too late. The stampede to turn grasslands into wheatfields, via the mechanized ploughing of thousands of hectares, in a region where the wind blew hard constantly and rainfall was unreliable, in order to boost grain production, created the infamous "Dust Bowl" that devastated the lives of farmers in Kansas, Oklahoma, Colorado, New Mexico and Texas. The topsoil was blown away by the incessant dust storms and severe gully erosion developed on the hard, bare soil surface when it rained.

Such inappropriate land management practices were not confined to North America. In the late 1940s in British colonial East Africa, the eponymous groundnut scheme was a failed attempt by the post-war British Government to cultivate over 12,000 km^2 of its African trust territory, Tanganyika, with groundnuts. The aim was to increase margarine supplies in Britain and develop a neglected backwater of the British Empire. Despite an enormous effort and at a cost of £36 million, the project was a disastrous failure and was finally abandoned as unworkable in 1951. It failed because the environment, especially the clay soils and rainfall, were unsuitable for commercial groundnut production. The failure to match farming systems and crops to the soil and its surrounding ecosystem continued in the developed and developing world throughout the remainder of the 20th century. The urgency of feeding the burgeoning world population was seen as sufficient justification for converting tropical rainforest into oil palm plantations or cornfields despite the advancing scientific knowledge that such practices carried great risks to the soil itself and the wider environment.

It is only in the 21st century that the voice of science, spear-headed by the United Nations Food and Agriculture Organization and numerous national soil science societies, has been heard and some restorative action taken. But by 2022, a third of the planet's land was severely degraded and fertile soil was being lost at the rate of 24 billion tonnes a year according to the latest UN Global Land Outlook 2 Report (UN, 2022). Human damage to the planet's land is accelerating while half the world's people are suffering the impacts. The world's ability to feed a growing population is being put at risk by the rising damage, most of which is caused by intensive, mechanized monocultural food production.

Degraded land which has been depleted of natural resources, soil fertility, water, biodiversity, trees or native vegetation – is found all over the planet. Many people think of degraded land as arid desert, rainforests maimed by loggers or areas covered in urban sprawl, but it also includes "green" areas that are intensively farmed or stripped of natural vegetation. Growing food on degraded land becomes progressively harder as soils rapidly reach exhaustion and water resources are depleted. Degradation also contributes to the loss of plant and animal species and can exacerbate the climate crisis by reducing the Earth's ability to absorb and store carbon.

Without urgent action, degradation will spread further. By 2050, according to the United Nations (2022), an area the size of South America will be added to the toll if current rates of harm continue. Reversing this trend is key to establishing a sustainable environment, with a balanced healthy soil layer, that can feed the world's population. In order for the restoration of the world's soil and degraded land to succeed, the soil must be considered as an international asset that is protected by international and national laws so that soil and land degradation becomes an indictable offense. An International Soil Assessment Agency, probably under the UN will need to be established along with its national counterparts, that are the sole arbiters of soil and land degradation. Branches of this agency, with qualified staff, will be required at all government administrative levels.

But reclaiming degraded land can be a long and expensive business depending on how badly damaged the soil and landscape is. The damage will depend on specific soil and broader environmental factors. An understanding of the mechanisms that cause soil and land degradation is essential for remedial measures to succeed.

4.5.2 Soil Erosion

The rapid erosion of soil by wind and water has been a problem since man began cultivating the land (Morgan, 2009). The factors that influence the rate of erosion are rainfall, runoff, wind, soil type, slope, plant cover and the presence or absence off conservation measures. These factors are grouped under the headings of energy, resistance and protection. Energy represents the potential ability of rainfall, runoff and wind to cause erosion, termed the erosivity. The resistance factors affect the power of the erosive agents; thus, the construction of terraces and wind breaks reduces the length of runoff and wind blow, respectively. The erodibility of the soil will depend upon its mechanical and chemical properties; factors like terraces that encourage the infiltration of water, thereby reducing runoff, decrease erodibility, whilst any activity that pulverizes the soil like mechanization, increases it.

TABLE 4.6 Common erosion gully types

	GULLY TYPES	*TYPICAL MATERIALS*	*DESCRIPTION*
A	Axial gullying	Gravelly deposits	Consists of individual gullies with single head-cuts that retreat upslope by surface erosion.
B	Digitate gullying	Clay loams	Retreat occurs in several head-cuts extending in the direction of tributary depressions.
C	Frontal	Loamy sands with columnar structure	Collapsed piping long riverbanks.

The effects of an extreme rainfall event can be very dramatic. Hudson (2015) working on soil erosion studies in Zimbabwe, found that 50% of annual soil loss occurs in only two storms and that, in 1 year, 75% of the erosion took place in 10 minutes. Soil erosion from rainfall is caused by rain splash, overland flow (sheet erosion), rill erosion, gully erosion and mass movements such as land slips of saturated soil on sloping ground. The mechanisms act as a progression determined by the intensity and duration of the rainfall event, the physical nature of the soil, the slope of the land and the vegetative cover. The momentum of a raindrop is transferred in full to the soil surface where it acts as both an agent of consolidation and dispersion. The consolidation effect is best seen in the formation of a surface crust, usually only a few millimetres thick, which results from the clogging of the pores by soil compaction (Morgan, 2009). The most important effect of a surface crust is to reduce infiltration capacity and thereby promote greater surface runoff and sheet erosion. Surface sealing or "crustability" decreases with increasing contents of clay and organic matter since these provide greater strength to the soil; thus loams and sandy loam soils are the most vulnerable to crust formation.

Gully erosion is both spectacular and complex. Gullies are relatively permanent steep-sided water courses, characterized by a head-cut and various steps or knick points along their course, which experience ephemeral flows during rainstorms. Gullies are almost always associated with accelerated erosion and landscape instability. Three types of gullying, related to differences in soils, are recognized (Table 4.6).

4.5.3 Reclaiming Degraded Land and Soil

The first step is to determine the ecological parameters within which the parcel of degraded soil and land sits. Establish the regional climate type and the physical properties of the soil and land including slope, geomorphology, land use, hydrology and natural vegetation. Then establish the actual status of degradation – the reality check – the type, percentage and degree of soil/land degradation in order to determine the most appropriate remedial treatment required. For example, degraded land within an area of commercial monoculture such as in an oil palm plantation or large US corn farm, can be restored on a piece-meal basis but will remain as part of the larger monoculture ecosystem in the short to medium-term. Remedial action would be endorsed by the plantation owners to restore the degraded land and protect its further spread, with the possible legal consequences. It is essential that soil experts and ecologists are included in the remedial action plan. This should be a legal requirement.

Soil and land degradation mapping should be undertaken by the empowered government agencies at all administrative division levels; for example, this could be province, district, sub-district, village and hamlet.

4.5.4 Soil Remedial and Restorative Measures

There are three classes of soil/land remedial and restorative measures, namely, physical, chemical and biological. Generally, a combination of all three is required to improve degraded land. Physical methods include ripping, ploughing, tilling and harrowing, all of which aerate the soil and break down harmful soil crusts and compacted subsoils. The most extreme physical method of soil remediation is the actual removal of the soil layer and its replacement by a fresh layer. This only occurs in cases of soil contamination by industrial effluents containing high levels of hydrocarbons, organophosphates and heavy metals in industrial sites close to residential areas. In such cases, there are other remedial options including containment, solidification, vitrification, soil washing, soil flushing, pyrometallurgy, electrokinetic and phytoremediation (Chen et al., 2016).

Containment measures are often carried out as measures to prevent or significantly reduce the migration of contaminants in soil or groundwater. *Solidification/stabilization* using cement is an alternative waste treatment with the aim of reducing the environmental pollution. *Vitrification* is a remediation technique in which a substance is transformed into glassy materials through the application of heat at high temperatures, with the addition of additives, whereby the soil/waste melts and a stable glass is formed. The resulting glass and glass-ceramic products have good mechanical and chemical resistance, immobilize the contaminants and prevent further negative environmental impacts occurring. *Soil washing* is a remediation technique based on the theory that prone contaminants bind to fine-grained soils (silt and loam), which tend to bind to coarse-grained soils. These techniques include ex situ remediation, which involves removing harmful pollutants from the soil by washing the soil with liquid, scrubbing the soil and then separating the clean soil from contaminated soil and washing water (Dellisanti et al., 2009). *Soil flushing* is an in situ remediation technology that can occur in an unsaturated zone, a saturated zone or both. The flushing solution increases the mobility or solubility of the adsorbed contaminants to the soil matrix. *Pyrometallurgy* is a process of metallurgical extraction using heat energy with temperatures reaching up to 2,000°C. *Electrokinetic* remediation is a technology for restoring soil contaminated with heavy metals and organic compounds through an in situ process using low voltage and constant direct current.

Phytoremediation is a technique to use plants and their parts to decontaminate waste and environmental pollution problems either ex situ using artificial ponds or reactors, or in situ or directly in the field on soil or areas contaminated with waste. Phytoremediation is also characterized as a form of contamination retention that is intervened by non-food plants, including trees, aquatic plants and grasses.

Chemical measures include the application of calcium carbonate ($CaCO_3$), also known as liming, and gypsum ($CaSO_4$), both of which adjust the soil acidity (pH) towards the more favourable neutral condition of pH 6.5–7.0 for most plants, and improves soil structure, aeration and stability. Biological methods include the application of animal manure to the soil, the application of green manure in the form of residual plant material, mulching and the planting of leguminous cover crops that protect the soil from raindrop impact and provide nitrogen and available phosphorus to the soil.

Which method or combination of methods are adopted will be predicated by the soil and land degradation assessment. In any area, the degradation assessment is likely to reveal a number of problems that require action over a timescale varying from the immediate, through the short-term and medium-term, to the long-term such as the restoration of saline or sodic soils using a combination of soil flushing and phytoremediation over several years.

4.6 SUMMARY

Most of the natural events described above are potential disruptions to development with the larger ones being significant. It reminds us that the Earth is constantly changing, although mostly in geological time, which generally should allow us to make the necessary adjustments. This was exemplified in the southeast of England, where up to about 10,000 years ago; the Dogger Bank (Dogger Land) off the English east coast was above ground and people lived there. However, the Bank gradually disappeared under the sea between 7,000 and 10,000 years ago, due to a combination of erosion, land sinking, and rising sea levels. The nearby coastline of East Anglia is also now eroding in bites from sea erosion. As a pre-emptive measure to counter increased flood risk to London, a barrage was built across the Thames estuary a few decades ago to protect the city from being inundated with water from combinations of high tides and storm surges. This measure has been, so far, an obvious success.

The possibly more troublesome changes and challenges in the shorter term are those that we are witnessing due to human-induced changes to the climate and how we come to terms with the continuing increase in human numbers since the early 20th century.

NOTES

1 Actually, the global average air temperature has reached this figure already, 2024.

2 Cenozoic meaning 'recent life', from 66 million years ago until the present day. It is characterized by the dominance of mammals, birds and flowering plants, a cooling and drying climate and the current configuration of the continents. The Cenozoic era is split into three periods: Paleogene, Neogene and Quaternary.

REFERENCES

ADB (2011) *Guidelines for climate Proofing Investment in the Transport Sector Road Infrastructure Projects*. Asian Development Bank: Manila. Available at: https://www.adb.org/sites/default/files/institutional-document/32772/files/guidelines-climate-proofing-roads.pdf

Beck, C., Zimmerman, N. E. and Wood, E. (2018) *Present and Future Köppen-Geiger Climate Classification Maps at 1-km Resolution*. Available at: www.nature.com/scientificdata

Beggan, C. (2025) On the move. *Geoscientist* 35(2) 25–29.

Chen, J. P., Wang, L. K., Wang, M. H. S., Hung, Y. T., and Shammas, N. K. (2016) *Remediation of Heavy Metals in the Environment*. CRC Press; Abingdon.

Cook, J. R. (2022) Reducing future climate impact on rural transport infrastructure in developing countries; the role of engineering geology. *Quarterly Journal of Engineering Geology and Hydrogeology*. Thematic collection: Climate change and resilience in engineering geology and hydrogeology. Available at: https://www.lyellcollection.org/cc/climate-change-and-resilience-in-engineering-geology-and-hydrogeology

Dellisanti, F., Rossi, P. L., and Valdre, G. (2009) In-field remediation of tons of heavy metal-rich waste by Joule heating vitrification. *International Journal of Mineral Processing* 93(3):239–245.

Geiger, J. (1961) Revised new edition of Geiger, R.: Köppen Geiger, *Climate of the Earth*. (Wall map 1:16 million). Klett, Perthes: Gotha.

Hardin, E. (2025) *What will happen when the Earth's poles flip*? Available from: https://www.ncesc.com/geographic-faq/what-will-happen-when-the-earths-poles-flip

Hudson, N. (2015) *Soil Conservation*, 3rd Edition. New India Publishing Agency: New Delhi.

IPCC (2014) *Climate Change 2014: Synthesis Report. Contribution of Working Groups I, II and III to the Fifth Assessment Report of the Intergovernmental Panel on Climate Change*. IPCC, Geneva, Switzerland.

IPCC (2021) *Climate Change 6th Synthesis Report*. A report of the Intergovernmental Panel on Climate Change. Available at: https://www.ipcc.ch/assessment-report/ar6/

Köppen, W. (1936) Das geographische system der klimate. In *Handbuch der Klimatologie* (Köppen, W. and Geiger, G. (eds.)). Gerbruder Bontraeger: Berlin, Germany (in German).

Kottek, M., Grieser, J., Beck, C. and Rudolf, B. (2006) World map of the Köppen-Geiger climate classification updated. *Meteorologische Zeitschrift* 15(3):259–263.

McGuire, B. (2022) *Hothouse Earth: An Inhabitant's Guide*. Icon Books; London.

Morgan, R. P. C. (2009) *Soil Erosion and Conservation*. John Wiley & Sons; Malden, MA.

Richter, C. F. (1935) An instrumental earthquake magnitude scale. *Bulletin of the Seismological Society of America*, 25 (1):1–32.

Ritchie, H. (2025) *Who Emits the Most CO_2 Today?* OurWorldinData.org. Available at: https://ourworldindata.org/annual-co2-emissions

Santer, B. D. et al. (2013) Human and natural influences on the changing thermal structure of the atmosphere. *Proceedings of the National Academy of Sciences* 110:17235–17240.

UN (2022) *Global Land Outlook 2. Land Restoration for Recovery and Resilience.* United Nations.

USDA (2017) *Soil Survey Manual.* Handbook No. 18. Natural Resources Conservation Service. Soil Science Division: Washington, DC.USGCRP (2017) *Climate Science Special Report: Fourth National Climate Assessment, Volume I.* U.S. Global Change Research Program: Washington, DC.

Whitten, D. G. A. and Brooks, J. R. V. (1972) *The Penguin Dictionary of Geology.* Penguin Books, London.

Wikipedia (2025) *Tectonic Plates.* Available at: https://en.wikipedia.org/wiki/Plate_tectonics

World Bank (2024) *The World Bank Climate Change Knowledge Portal.* Available at: https://climateknowledgeportal.worldbank.org/

The Environment 5

Scott Younger, David Parry, and Mark Murray

5.1 INTRODUCTION

Humans arguably have the biggest impact of all species that live in the environment of planet Earth, especially when their numbers have multiplied significantly. As stated previously, the population of the world has expanded almost three-fold over the past 60 plus years, such that it stands at over 8.1 billion today, in 2025, and is forecast to be about 10.9 billion mid-century. The pace of growth has slowed down and is slowing down further such that the population, after reaching about 12 billion, is expected to start declining in the last couple of decades of the century. The population is forecast to be about 11 billion or a little under by 2100 (UN, 2024).

Apart from significant population growth since the 1960s, there has also been a number of important developments over the same period. These are the results of human endeavour, as pointed out before – significant developments of the computer, television, air travel and advances in scientific and medical research and technology with the faster spreading of knowledge and its practical facilitation. To this has been added, comparatively recently on the back of computer developments, Artificial Intelligence (AI) (Kirwan, 2024). Even in the next decade or two, there are some surprising developments coming up.

A main issue that has started already is going to be changes in the distribution of the populations in Asia and Africa. The Chinese population, already declining from a peak of approximately 1.42 billion, the largest in the world until a few months ago, is forecast to be approximately 800,000, a change of over 600,000, by the end of the century. India, past its peak of over 1,400,000 today, will also by mid-century, be on a downward path. The major expansion, by contrast, will come about in sub-Saharan Africa which is forecast to increase from today's 1.4 to 4.2 billion by the end of the century (UN, 2024). Over three quarters of the population of the world will then reside in Asia and Africa, a large proportion in the equatorial zone and a little beyond.

There will be little change for the rest of the century in Europe or America from today, nor in mature Asian countries such as Japan and South Korea. Most of

DOI: 10.1201/9781003566274-5

the additional population over the remainder of this century will basically come from African countries. It is becoming evident now as much of the migration across the world takes place from these countries to Europe.

Of the total area the planet provides on its surface, just under 30% appears above the oceans as land, where all humans dwell with other mammals, birds, seals and reptiles and some amphibians. The rest of life, whales, sharks, seals, etc., and, of course, fish, are primarily sea creatures. The oceans, appearing at some 70% of the planet surface, are relatively unexplored, and there is much for marine scientists to still look into, not only in terms of resources.

The fish provide a very important food source, vital for many communities as well as larger ocean predators. Our weather is linked to what happens over the oceans, e.g. the so-called trade winds, and thus great attention has to be paid to what is happening there with regard to climate change and changes that we effect, for instance by overfishing. Ritchie (2024) with a comprehensive analysis of available data, however, has shown that, while there might be concern, there is no danger of fish stocks drying up in the foreseeable future, provided suitable care is taken.

Along with the population growth since 1960, in the past couple of centuries, mankind has been gradually denuding the planet of its forests and increasing the space created for agriculture. The global forest area has declined by 60% since 1960 (Bennet, 2022) and one-third of arable land has also been depleted by soil erosion or degradation (WWF, 2025). The stripping of forested land has had a negative effect on the climate by also increasing carbon emissions to the atmosphere. Trees of the forest were formerly absorbing a significant amount of carbon emissions, but now there is only 15% of forested land left. A similar story can be found more recently right across the world such that the Amazon rainforest, called the lungs of the world, is now in a 'perilous' situation. Similarly, the island of Borneo, which has lost much of its forest to grow oil palm and has reduced the habitat of most of the species, e.g. the orangutan (man of the forest), who depends on the trees. These examples make up much of the natural forest loss referred to above (Bennet, 2022).

In order to take action on the world's climate by putting a stop to the increase in carbon emissions, much of the attention has been on the use of fossil fuels, and the leadership to largely eliminate their use by the end of the century is being taken by the UN. They are trying to work with national governments, not always with much heed being paid, especially by the countries who are responsible for most of the carbon emissions! However, there is evidence of some serious intent in a number of places. The apparent alarming warming of the climate was the main item on the agenda at the UN COP 21 meeting in Paris, 2015, where all governments of the world were warned of the disastrous consequences if they did not act immediately to significantly reduce their contribution to carbon emissions.

More recently, attention has been drawn to use of groundwater which is the basis of most of the freshwater that is used. A fuller discussion on the overuse of this vital resource is given in Chapter 7. Suffice to say here that this requires more attention from now on, for instance the cleaning of 'dirty' water and its reuse, to avoid a crisis, an unsustainable situation arising in the next few decades.

5.2 THE EXISTING URBAN ENVIRONMENT

Of the total population of the world today, 55% reside in an urban environment. By mid-century, the total population is forecast to be approximately 10.5 billion, with a significant increase expected to dwell in towns and cities; that is 68% of the population (UN, 2024), or approximately 6.9 billion by the next two and a half decades. The urban population will thus increase by 2.5 billion.

The urban population of the UK as of 2024 is given as 84.4%, much the same as for western Europe. That of USA is given as 82.4%, almost the same.

There is thus a significant difference in comparison with the undeveloped world. Nearly all the growth in population will take place in undeveloped Africa, as indicated in Figure 1.2. There will be pressure in Africa to build homes as quickly as possible sometimes, and hopefully certainly in time, prioritizing 'green' materials. However, steady growth can be expected of small communities evolving into towns giving way to larger towns and some new cities from the more favourable towns. Nevertheless, given the rate at which the population is expanding, we will see initially, and for a few decades, many sizable communities still dwelling in tented or shanty accommodation till suitable, more solid housing is provided.

The rapid increase in African population is the principal driver of emigration to Europe in the past few years to escape from poverty, or fleeing strife in places such as Eritrea, Sudan, Rwanda and Congo, in the hope of making a better life for themselves.

There are over 4 million urban centres in the world from small towns of just a few thousand up to a few megacities and even meta-cities (over 10,000,000 residents). Many of the largest cities have grown up in Asia in the past half century: Bangkok in Thailand, Beijing and Shanghai in China, Jakarta (Younger et al., 2015) in Indonesia, and Delhi and Mumbai in India, for example.

Jakarta is only 120 km distant from Bandung, a city in the hills to the south and the capital of West Java, now connected by toll roads as well as a 'fast' train to Jakarta, an upgrade from the old colonial railway. It is forecast that greater Bandung, which has also been expanding quickly, will merge with Jakarta which continues to spread eastwards, along the north Java coast, by the next decade or so. This will make one conurbation of 70 million inhabitants building out from the few stopping points along the railway route where there is already a community in existence. It provides some opportunity to use materials and techniques for buildings that reduce carbon emissions.

The implications of the substantial increase in population numbers since 1960 and commensurate rate of growth, in general and in cities, in particular, have not been realized by many governments across the world. There is a shortfall in accommodation in several countries, which raises a housing problem. The new government in the UK has recognized this, somewhat belatedly, and put new building as one of their top priorities. The problem is not just in the UK but also arises in similar countries. For instance, in Europe, unscheduled immigration adds to the problem, although there is now a decline in recent times in the indigenous population which can only be replaced by inward migration. Thus, the changes will show quite a significant difference in the racial mix by 2100 compared with a century earlier.

The main attention will have to be towards developing Africa, however, not just in significant professional/technical help in various disciplines, but also in governance. This is a big coordinating project for the UN in harmony with larger, wealthier and more mature nations as well as the World Bank, ADB, AfDB and other multilateral agencies. Solutions are different, varying from country to country, and the need for appropriate engagement is required as soon as possible.

Construction in appropriate locations on suitable land so as not to spoil the environment is important, and to use modern techniques, and the best materials that are available with energy efficiency during operation must be understood. Consideration has also to be given to any effects in the future from climate changes, different locations being checked against, for instance, the climate zones suggested by Koppen-Geiger (see Chapter 4).

Many of the major cities in the world are located near or at a shoreline of a country, and many of them have grown up at the mouth of an important river with easy access to the sea to provide linkages to trade and services from and to the country's interior. This applies to most of those cities which are mentioned above and many others and reflects our need to develop at locations that provide convenient access.

In some parts of the world, land, unburdened by thousands of metres of snow and ice as a result of millennia of changing climate, gradually heating, is slowly rising. This has happened in northern Britain in the past, for instance, and consequences of this can now be very accurately calculated from satellite-based techniques revolutionizing geodesy, making it now possible to study both the uplift and subsidence of land masses in 3D (Baggeet al., 2021). Other parts of the world have disappeared under the sea, like 'Dogger Land' between England and mainland Europe between around 5,000 and 7,000 years ago, due to tectonic movements balancing sea-level rise and uplift of land masses due to melting and removal of ice sheets (Cunliffe, 2012). Similarly, during the last 2 million years, the area in Southeast Asia known as 'Sundaland' (Bali, Borneo, Java and Sumatra in Indonesia, and the Malay Peninsula) varied from being an uplifted land mass to having a sea level 3–4 m higher than now depending on how much oceanic water was locked up in ice masses (Mann et al., 2019).

The land area of the world is not stable in a macro sense, although within a period of several centuries or more, it can be perceived as stable. It is subject to movements generated by adjustments of the seven or eight major tectonic plates and erosion by the weather, dictated by the climate. In terms of sustainable development in a short- or medium-term time frame, say one or two centuries compared with the long term of a million and more years, on living on the planet, we can only sensibly plan for the relatively short term. However, we must be aware of and take such actions deemed necessary towards mitigating consequences of major events, such as life-threatening earthquakes and volcanic eruptions. This applies mostly to rural land areas, but also to towns which are located in hilly areas, particularly where there is the possibility of earthquakes or unprecedented flooding. A few countries, such as the Netherlands, largely lie below sea level and are protected by a seawall or dyke, which needs periodic monitoring.

With that as background, there are a number of matters that we can do to make sure that a structure that is designed is stable and fit for purpose within the sustainable period considered. There are several important cities on river mouths/estuaries, which have been built on soft, still consolidating ground, which are sinking. Sometimes, the cities are built on flood plains and are prone to flooding, especially in extreme wet conditions, which are indicated to be more frequent as the climate changes.

Alternatively, the acting drainage paths, often the cities' streams and rivers, have to be attended to, perhaps widened, to allow faster release of upstream water. Some cities have done this, as these cities have grown substantially, particularly in the past 60 years, requiring additional drainage channels. Cities, which are located in inland areas and are prone to heavy rainfall, particularly if this is unprecedented, can show excessive damage, e.g. landslides, and loss of life from flooding.

Two matters which have raised some concern in Europe and Asia relate to sewage and air pollution, particularly in densely populated cities, both of which require immediate or continuing attention.

Air pollution has been a problem for many centuries. The Greek and Roman civilizations burned wood for cooking and heating. Seneca (CE 65) wrote about it and was relieved when he could leave Rome and breathe the unpolluted air away from the city. In the past millennium, there have been several writers who complained about air pollution in cities, e.g. London. Although wood continued to be used, the use of coal, because of its greater energy efficiency over wood, and secondly the development of the steam engine in the latter part of the 18th century, led to wood being superseded by coal.

In UK in the latter part of the 19th century up until coal was banned for direct use in urban homes in the 20th century by The Clean Air Act of 1956 (Wikipedia, 2025), it was the main polluter of air, giving cities dense, distinctly unhealthy, filthy fogs particularly when the temperature was cold and the atmosphere still. The improvements in the air that is breathed in the last 60 years, primarily from phasing out coal and replacing it with less polluting gas over the first decades and then latterly from the gradual introduction of renewable energy sources, are considered to be a factor in improving the longevity of some of the population.

There has been a significant increase in vehicular traffic on the roads, particularly in cities, increasing the carbon emissions to the atmosphere, as well as making it unhealthy for residents, partially counteracting the benefit from the removal of coal. There is still much to be done in major cities where there are major traffic jams. In time, during this century, there is expected to be a substantial reduction in city-generated carbon emissions from the slow but steady introduction of electric vehicles (EVs).

Introducing energy-efficient materials in the large volume of building that has to take place in not only Africa and developing Asia, but also in the older existing settlements of Europe and America, will also play an important part in reducing carbon emissions. Reductions in carbon emissions that are deemed necessary to be achieved by 2100, to keep the average temperature rise to 1.5°C as agreed by all nations at the Paris UN Climate meeting, will require changes in the way we live, as well as the way in which bulk energy is produced and the point of its delivery. In fact, the 1.5°C limit was passed in mid-2024, and it would seem that fixing a figure on average world temperature is far too complicated to pursue as a limit for all the countries of the world, particularly when the biggest emitters of GHGs either are sceptical of the reasoning behind the target or have other short-term priorities. It is deemed more sensible to divide the world into climate zones, as discussed in Chapter 4.

5.2.1 Pollution

The sewers and sewage systems in developed countries have been found to have been mostly inadequate in the last few years with the significant increase in population as well as a required upgrading of the routine maintenance. This was found out in recent times in England with the seven area water management organizations, each responsible for all water-related activities for a given sector of the country, when the sewage infrastructure that each organization was responsible for, became seriously overstressed. The overspill of sewage was just dumped in rivers and still is to the detriment of the environment (several articles in The i-newspaper, 2024). The overspill was and is increasing at an alarming rate, and it was obvious to the general public that the capacity of the sewerage systems had to be increased. Not only was the old infrastructure totally inadequate and had to be upgraded, but extra capacity also had to be provided for an expanding population as well. The regulatory body and the management of the water companies had been found wanting!

The overspill going into the rivers, if continued unchecked, is dangerous for wildlife and the near-surface groundwater and ultimately the health of the population generally, and could lead to pandemics of cholera, etc. It has also been noticed that pesticides and other sprays that are being used by farmers to keep down unwanted bugs have been collecting in the surface water and polluting the streams and rivers.

Plastics generally and electronic waste, in particular, have also been demanding attention, rather than just being dumped in landfills and leaving a problem for future generations, or shipping off to other largely unaware countries or gathering and polluting at sea, where the small particles can be digested by fish. Solutions are being sought to make these materials reusable by first crushing their ingredient parts to a minute (molecular) particle size and thereafter producing recycled materials for reuse (Scindo, 2024).

Cities are known as hot sinks. Thus, all measures to reduce the impact of carbon emissions should be welcomed. The main ones are thought to have been aired and must be a factor in whatever future activity. There are a number of ideas that have come forward in the past two or three decades. The following discussion looks at what is required to be done for existing urban infrastructure, for today and future generations in terms of what they need and want.

An important consideration in any development concerns what should be done in the urban areas about trees, where far too many have been cut down and not replaced. The significant loss of trees in urban areas has meant more carbon has had to be emitted to the atmosphere, an unsustainable position as more cities have been built and have been growing in size.

This has been realized in the past three decades and replanting is being undertaken, perhaps not fast enough. The planting of trees has to be nurtured as urban living expands. It is a reminder that urban forests are useful for capturing essential carbon gases from which they produce oxygen and play a useful role in human lives and the life of the planet. Today's settlements, whilst probably affecting the 'smart green city' agenda, still appeal to the market through urban sprawl, justifying its open, healthy, private garden spaces.

5.2.2 The 'Smart City'

The term 'smart city' is becoming more common as a means of attracting all types of people, perhaps especially professionals. It means that a city is quick to take opportunities and provide forward-thinking strategies to create designs for comfortable living environments. IBM (2023) defines a smart city as an urban city where technological and data collection help to improve quality of life as well as the sustainability and efficiency of city operations. Kirwan (2024) writes about this and extends the argument to AI.

Cities in Asia that take the accolade of smart city seriously are, for instance, Singapore, Huawei and Bangkok and a growing number of others in Southeast Asia. Further, retrofitting services and utilities, while difficult and probably expensive, must be considered in the light of environmental sustainability goals, and not forgetting decarbonization aims.

It is prudent to take a survey of the green spaces as well as the surrounding trees that exist to finalize the immediate needs of existing citizens, their health and well-being, noting up-to-date thinking on how to balance shelter with open ground.

Coping with the transport imperatives is a major issue in cities. After World War II for about a decade, the majority of citizens did not have cars, and relied on public transport to move around, in those days mostly by trams or buses. While there was a gradual increase in both public transport and private cars through until the 1960s – trolley buses adding to the public transport list in the late 1950s – the real take off in means of transport was towards the later 1960s, when private motor cars began to proliferate. The next 30 years, in particular, saw rapid road building in towns and cities as well as cross country where fast motorway connections were put in between the main cities. Nearly all these means of transport were powered by fossil fuels, petroleum and gas.

We must plan ahead for years to come, including for opportunities from the internet of things (IoT) and the many challenges that are expected from AI. They will undoubtedly change the scope of lives – smart cities.

For citizens, comfortable living, gathering places, squares will have to be created to give a sense of community togetherness with neighbourhood centres providing amenities within a 10–20 minutes-walk for residents. If there is an existing road or track to be rearranged or simply a new replacement secondary road to put in, it should be designed to cater for pedestrian, cycle, scooter or other form of environmentally (and healthy) means of travel, such as autonomous driverless delivery services by EVs. These will gradually become more commonplace in the future and have already been introduced in some cities.

In Southeast Asia, the self-determination of countries really did not get properly underway until the 1970s. The forms of governance were being resolved and citizens, with some experience, were found to hold the offices of state. It took a decade or more for these new nations to settle down in the business of government. The harmony that resulted from decades of peace has made the countries that encompass the Association of Southeast Asian Nations (ASEAN) a formidable trading bloc – the one exception being Myanmar which is experiencing self-imposed civil war, as consequence of a military coup.

The recent 'working from home' diktat, a result of the COVID pandemic was quite significant and considered quite successful such that there is planning for an increase in home working, which would mean a revision of the requirement for office block space and home space and possibly a redesign of space generally for the future. The principle of working from home, however, does not suit some operations, and the system is still under trial. In the meantime, through adaptive reuse, offices which find the system workable can convert the space for reuse as apartments or educational or laboratory facilities, for example.

Sometimes the citizens of urban centres are guilty of polluting the (largely developed) world's rivers and streams, and estuaries. These rivers provide vital water for the foundations of trees, shrubs and flowers, their roots. However, these are quite often mixed up with plastic and other human waste, flotsam and detritus! Harmful agricultural by-products, such as pesticides and animal droppings, medicines and drugs, etc. are also to be found, and some of these are dangerous.

5.2.3 Vital Urban Resources – Water and Energy

Water and energy are two vital resources that are needed for survival, and these will be covered in greater depth in Chapters 7 and 6, respectively. For urban areas, quite a number of towns and cities have a full range of the level of development of water resources, the exception being in terms of development of sanitation which is well behind.

Very large cities may be served from more than one source of water supply. At the most mature level, drinkable clean water on tap is available every hour each day, as in the UK. The step down in provision is clean water but not treated for drinking quality. Then when there is a limited supply, water is provided for given hours each day. Such dwellings are in areas supplied in bulk from the resource water treatment works to a sub-centre from where the water is distributed to different households. Too many households in developing countries, however, still have to gather their daily needs by a visit to a water tank, which is filled up periodically.

Where there is no easy access to water, rainwater is sometimes harvested and stored in similar large tanks, and sometimes there is a combination of two of the methods. However, there are still too many countries that are desperately short of water, a situation which is exacerbated by the large increase in population. It is trite but true that no water no life! From now on, we are going to have to be much more efficient in the way that we use the resource because of overdrawing the groundwater, as discussed further in Chapter 7.

Sewage requires urgent attention, especially in urban environments. The current situation suggests that many places in the world, urban areas in particular, still do not have sanitation, adequate or otherwise. Operations for sewage treatment have fallen badly behind, especially with the rate at which the population has increased in the past half century.

For instance, there are regular reports of the main rivers in England being polluted by discharges from overspills of raw sewage, and the private sector water companies, who are responsible for dealing with this, and the regulators for the water industry, being rightly criticized in the press, and active private environmental groups There is much work to be done, probably for decades to come in many countries. Neglecting this task opens up the risk of dangerous pandemics.

The scope of energy supplies has widened in the last 50 years, especially since attention has been paid to reducing significantly the use of fossil fuels, particularly for servicing urban areas. Reference is made, in the first place, to the work of Koppen–Geiger (Chapter 4) to establish the particular climate for a given area and then decipher the most economic but reliable energy source. Only certain locations are good for using offshore wind, like UK and Norway and along the eastern Atlantic seaboard. Alternatively, many countries in or near to equatorial climates may consider the sun as a prime source for their use. However, not all countries, notably some of the largest, accept the UN's advice on climate change and make use of fossil fuels extensively or have delayed their transition to renewable energy for development reasons.

5.2.4 New Urban Settlements

Looking to the future, subject to national or even international master-planning imperatives, while the human population continues to grow, there must be today, instead of the current tendency for urban sprawl, a plan to accommodate this in appropriate locations set aside until the population stops growing. As stated, this is forecast in about 50 years' time when the population is expected to slowly decline. The gradually expanding population in European countries is today comprised largely of new immigrants, since there is a small but gradual decline in the aging indigenous populations.

There is a plan for 'Edge Cities' to limit or reduce the size of existing settlements. These are situated more than 30–40 km and more away from existing overbuilt urban centres, suffering from overspill. There will be a continued demand in China and India, for instance, for huge numbers of their populace, which are not yet housed properly, which could be accommodated in this manner.

These unassimilated settlements or New Towns which, as the names imply, are new habitats geographically distant from existing ones. These are seen as an opportunity to arrive at socio-economic–environmental urbanization solutions, arrived at with full stakeholder consideration, within a National Plan Framework. In ensuring that the planners and developers are in agreement, it is that if the "goals of development are consistent with the prevailing lifestyles and consumer demands of local (and incoming) residents, the applicability of bottom-up strategies will probably be higher than in a situation where these lifestyles and consumption patterns are thought to be a main part of the very problem. The decision processes must also be democratic and legitimate" (Brundtland Commission, Rio de Janeiro, UN, 1987).

With Scandinavian homes now twice the size of 35 years ago at an average of 50 m^2, but with contemporary, affluent Japanese living in 70% less, the question arises as to the environmental consequences when and if the populations of China, India and Indonesia look for the same standard of living (Naess, 2001). Time and income pressures, however, as well as location, would suggest that the 'Japanese model' would hold sway.

There is one further problem, however. Will there be smart enough strategies to ensure an adequate funding stream for the implementation of each project according to defined, sustainable environmental timetables?

The impacts on the environment from human imprints will vary according to the size of the problem, geography, what are seen as political imperatives, commitment to development sustainability, the market and available finance and unfortunately corruption in some form. It is not possible to address the consistent presence of corruption in planning new settlements when profits head the modus operandi priorities. A comprehensive coverage of corruption, how to deal with it, and its impact on development is given in Chapter 2.

Education is needed to gain an understanding in the processes that affect harmony between all life on Earth, thinking beyond individual successes measured through monetary wealth to the well-being of communities as a whole. The largest conurbations and most glamourized today tend to be new national capitals. Most will set off with laudable environmental aims which, to be effective, have to be credibly monitored. Ambitious future areas/cities include the following:

1. The 170 km-long by 200 km-wide 'The Line' in Saudi Arabia: a 'model' for nature preservation and enhanced liveability.
2. 'Telosa' on an undisclosed US desert (the most sustainable city in the world).
3. 'Biodiversity', on artificial islands, Penang, Malaysia (no cars and the use of recycled materials with ecological buffers).
4. 'Smart Forest City', Mexico: a botanical garden within a heritage area, an ecosystem where nature and city are intertwined and act as a single organism.
5. New capital city, Egypt in the desert.
6. 'Amaravati', Andra Pradesh, India: 60% greenery and water, bringing together decades-long research into sustainable cities.
7. 'Nusantara', Kalimantan, Borneo, Indonesia, the new capital city: it should be a shining example of what green and sustainable development should look like.

It can be seen that there are several countries of the world making or thinking of making sustainable green cities or simply large areas of land development. These plans would incorporate local skills, infrastructure and pioneering research. In addition, with best practice digital infrastructure, it is intended there should be enough to attract capital investment, to innovate new technologies – leveraging 5G, big data and IT, sensors, geospatial and blockchain, all covered by cybersecurity. This is what Kirwan (2024) calls in precise parlance 'Infostructure'; bringing together today's building knowledge with tomorrow's AI to develop a smart city.

For new cities planned and built on essentially 'green field' locations (some will incorporate existing small settlements), armed with today's environmental awareness, knowledge and ambition, this will ensure maximum environmental sustainability, with the aim of a more sustainable, zero-carbon human impact wherever the sites are located. Each would be planned in such a manner as would meet the current social, economic and environmental guidelines.

When development sites are selected, indigenous populations are not often meaningfully consulted in the usurpation of their traditional homelands and destruction of their culture and very existence. This is not deemed right, and consultation should be mandatory.

In the past, because of convenience whereby it is easier to build on flat ground than on hilly terrain, too often mature cities and new towns have been located on flood plains and peat lands which have to be drained, and afterwards several levels required because of land use change.

The decisions to replace the existing biome, such as primary and secondary forests, plantations, farmland, nature reserves and recreation areas, to accommodate new human settlements are too often expediently taken with sustainable promises. However, these promises are later superseded by other imperatives, e.g. boosting national identity, overpopulation. Issues increase ambient temperature from those of a 'green field' condition by 20°C–40°C and by 40°C–60°C in cities in the tropical conditions that are found in equatorial Indonesia, for example. No increase, however, is perfectly attainable through initially planting appropriate trees, incorporating such as white/'green' roofing, and designing for grey streets, adjacent water bodies and generous, interlinked green spaces; that is, building Kirwan's smart city (Kirwan, 2024).

5.2.5 Summary Observations

There are quite a number of issues that already exist and can be expected to exercise city governments for the rest of this and beyond century. They should be expected to deal with changes as they progress. Urban life will change, however, as the latest development of the computer age, AI, takes hold and significantly alters the workplace forever.

The two vital areas that require immediate attention concern the supply and use of energy and water, both of which will need substantial work and financial resources throughout this century and well into next. In the case of energy and water with the added issue of pollution, which can have a deleterious effect on the environment, there is no lack of resources although, in some parts of the world, depending on region there is a shortage, sometimes serious. The chapters that follow discuss energy and water more fully.

The urban environment relies on how the land is treated in terms of buildings, their use and care, roads, the linking through roads and main city serving public transport to feeder and minor roads, and railways serving the city's needs and providing links to other major cities.

What is being appreciated even more so than in the past is how much green space is needed, where and the value of trees, not just for providing space to give fresh air but also to absorb some GHGs. This has become especially important, as the population has expanded in the past 60 years and the GHG's given off by a greater number of motor vehicles has also grown.

5.3 THE SUSTAINABLE RURAL ENVIRONMENT

5.3.1 Background

In general, a rural area or a countryside is a geographic area that is located outside towns and cities. Typical rural areas have a low population density and small settlements. The Organization for Economic Co-operation and Development (OECD) defines a predominantly rural region as having more than 50% of the population living in rural communities where a 'rural community' has a population density of less than 150 people/km^2. Agricultural areas and areas with forestry are typically described as rural, as well as other areas lacking substantial development. Rural areas have unique economic and social dynamics due to their relationship with land-based industry such as agriculture, forestry and resource extraction.

Rural economics can be subject to boom-and-bust cycles and vulnerable to extreme weather and natural disasters, such as droughts. These dynamics alongside larger economic forces, encouraging urbanization, have led to significant demographic declines, called rural flight, where economic incentives encourage younger populations to go to cities for access to education and jobs, leaving older, less educated and less wealthy populations in the rural areas. Slower economic development results in poorer, less developed services, for instance, as in healthcare, education and infrastructure. This cycle of poverty contributes towards three quarters of the global impoverished living in rural areas according to the FAO, the majority of which are located in developing countries.

The start of the Industrial Age and the years that followed saw a major shift in the balance of population that lived in the rural community versus that which resided in towns and cities. This mostly happened during the latter part of the 19th century, the Victorian Age, and more spectacularly across the world since the 1960s. By 2050, it is forecast that some 75%–80% will be urban residents, while it was 2007 before the number of residents in urban areas overtook the number living in rural. This was when the overall population of the world was approximately 5.5 billion, although this figure has to differentiate between the developed parts of the world and the undeveloped, there being a significant difference when comparing each. Rural parts are immensely important for the health of the population since they account for more than 95% of world food production and agriculture employs many hundreds of millions, especially in India, China, Central Asia, the USA, Canada and Brazil. This situation is likely to remain for the foreseeable future despite changes in the urban/rural population ratio.

5.3.2 World Food Supply

Theoretically there is no global food shortage because more than enough food is produced to feed everyone in the world. So much food is produced globally, yet one third (1.3 billion tons) is wasted according to the UN. Local food shortages are caused by local circumstances such as conflicts, droughts, floods and fire. Some 10% of the world's population is affected by hunger with more than 60% being women. Extreme conditions can be found today in Sudan, 12 million displaced because of civil war.

According to the World Food Programme, ripple effects of the COVID-19 pandemic and the war in Ukraine have contributed to one of the worst food crises in decades, with acute food insecurity affecting 200 million more people globally than in 2019 due to rising costs of food, fuel and fertilizer. But there are bigger problems on the horizon. As the global population passed 8 billion in early 2024 and is predicted to reach over 10.5 billion by 2050, farmers, governments and scientists face the challenge of increasing food production without exacerbating environmental degradation and the climate crisis, which itself contributes to food insecurity particularly in the global south. The UN projects that food production from plants and animals will need to increase 70% by 2050, compared with 2009, to meet increasing food demand.

But food production is already responsible for nearly a third of carbon emissions as well as 90% of deforestation around the world. We use half of the world's vegetative land for agriculture (Searchinger et al., 2019).

The truth and possible pending catastrophe is that the rural environment in most places is under significant threat both from an unprecedented increase in population in the past 60 years, which was largely unforeseen until the past 20 years or so, and the impact of this population increase on climate in parallel with related changes to the environment.

The space reserved for growing the food crops required would seem to have reached the planet's capacity, and there is a need to find new approaches to feeding, principally the human population, but also the rest of animal life living and dependent on the planet, not forgetting the flora. Many believe the limit of natural production is approaching from some food species such as wheat. Further improvements would require genetic engineering, a promising field but one with many detractors. This is slowing down the application of genetic engineering to food crops and will continue to do so for some time to come, although eventually a combination of common sense and necessity will prevail but not for a decade, at the very least.

The increasing incidences of hotter weather, heat waves, causing faster evaporation with changes in weather patterns, particularly rainfall, and earlier leaf drop with the consequences of failed crops, have to be dealt with. Understanding and improving the natural condition and resistance of trees and plants faced with changes more appropriate to drier climes now need to be planned for. This will push the boundaries of horticulture. For instance, there has been a realization that some more of the land has to be kept or left as meadow and not all of it should be like the English style of garden with its moisture-loving, unsustainable, high maintenance watering of lawns.

5.3.3 The Future for Sustainable Rural Development

The good news is that the rural environment encourages sustainability and those that live there have a skill set that is based on sustainability and is generally absent from their urban counterparts. Generations of farmers, smallholders and agricultural workers have produced food sustainably for the rest of the world but, as the FAO has observed, we seem to have reached or be very near the planet's capacity and producing more food from the same area of land will require new approaches to agriculture. Higher yields are still possible using genetic engineering, although there are many who doubt the wisdom of this approach as discussed.

A more acceptable system of increasing, in this case, rice yields, is known as the Rice Intensification System or SRI, a farming methodology that aims to increase the yield of rice while using fewer resources and reducing environmental impacts. SRI focuses on changing the management of plants, soil, water and nutrients to create a more productive and sustainable system of rice cultivation. Despite its success and the large increases in yield reported (from 30% to 100%), the adoption of SRI has been limited primarily due to a lack of awareness and available training. SRI has been proposed as a prime example of how agroecological approaches to farming can address what the *Applied Economics* journal (Varma, 2019) describes as the impending global crisis in rice. The concept of SRI, namely using less resources like water, fertilizer and pesticide, applies equally to other grain crops and to legumes. This very sustainable agroecological farming system needs to be greatly expanded in both the developed world and the developing world. It should go a long way to ensuring the world has adequate food supplies for the foreseeable future without dramatic or drastic changes to existing farming systems.

Unfortunately, the rural environment is fighting a losing battle against increased urban expansion and encroachment as more and more people leave the countryside for the towns and the superior job opportunities, and perceived rewards offered there. In developing countries, the rural areas are large and can absorb urban expansion. The situation in some developed countries is more serious, and there is a real possibility that rural areas that do not produce food will disappear entirely due to urban encroachment. There is a very urgent need for planners and administrators to set realistic boundaries to urban expansion, backed up by the force of the law.

The layout of rural areas and the land use is unlikely to change very much in the developed nations but, unfortunately, deforestation is likely to rise as forest and scrub land is converted to single crop plantations of oil palm and timber for paper production especially in South and Southeast Asia. The large island of Borneo is a perfect example, where oil palm has taken over large areas of the island from pristine rain forest, the 'home' of the orangutan. Given the huge socio-economic benefits from such plantations, the trend is likely to continue until saturation is reached and market forces make the necessary adjustments to minimize damage to farmers, smallholders, owners and investors.

There is a growing trend of thinking in parts of UK and also Europe for bringing the land to the natural state by what is called 'rewilding'. This can be admired, but only for a few weeks in the year, and emphasis should be, for example, on evergreen bushes, conifers and so on, and encouragement of insect life, e.g. bees, which could also help manicured gardens.

Today's attempts in reintroducing wildlife should focus, in principle, on native species. The resulting changes in the environment/habitat, for argument's sake, for restoring forest cover, are expected to bring a significant change to the landscape, affecting water bodies – rivers, lakes – and encouraging access and tourism. Some rural land is privately owned, as found everywhere around the world, which will slow down most efforts at development. Reform is needed on the rights pertaining to balancing the ownership and the rights of human and other life where these are in conflict. Harmonious solutions need to be found. It is well understood that this will take some time and usually involve political discussion and persuasion.

The concept of 'legal personhood' giving, for instance, a nation's rivers legal rights to 'exist, persist, maintain, and regenerate', as already adopted in Ecuador (Rio Altrato), Bangladesh (all rivers given specific rights), Quebec (R. Magpie) and in a similar manner, Bali. Bali's co-operative water management system, *subak*, where water is used to construct a complex, pulsed, artificial ecosystem, could well be copied elsewhere.

5.3.4 Additional Issues

Since so much of the land, approximately 98.5%, is primarily rural, this including mountains and deserts, for instance, and is uncultivable, it is important that it is treated equally with the urban. One role rural environment plays is the off-setting and balancing of hot zones which the urban areas of the world generate, as much by way of transport as in the use of buildings, homes, etc.

There is growing controversy over the millions of tonnes of plastics, which are derived from fossil fuels, collecting in the world's oceans. These are swallowed by fish unwittingly and no one knows the extent of harm to humans this may cause from long-term eating of fish. The fossil fuel lobby is strong, and there is dispute with those who want to see use of fossil fuels significantly reduced.

Pyrolysis destroys all plastics and can convert it to syngas and then combust it to create energy. At the moment, most plastic waste goes into landfills storing up e-waste, e.g. laptops, which is another large issue. A potential solution has been identified, which breaks down plastic polymers into valuable molecules, enabling the creation of new products. The solution is timely, if not essential (Scindo, 2024). According to the WHO, 53 million tons of e-waste are created annually with only 17% collected and recycled.

Finally, there is marked uncertainty over the issue of climate change, owing to the role that will be taken by mankind over the next seven decades. This is unclear at this stage, and therefore, it remains that it cannot be said what the environment will be like at the end of the century.

5.4 OCEANS, AIR AND SPACE

Since some 70% is covered by oceans, it would seem fairly obvious that reasonable knowledge should be acquired of the sea. Above the surface of the oceans, the complex nature of the climate, the winds, their direction and strength, is quite well documented, and this information is well referred to by shipping. Under the surface of the largely unexplored oceanic environment, there is a vast range of life from tiniest plankton to the largest whales. It is becoming increasingly clear that these oceanic life forms are also becoming increasingly impacted by recent pollution.

Non-degradable plastics are a particularly difficult problem. The World Bank (2025) reports that 11 million metric tons of plastic flowed into the oceans in 2016 and that this could easily become 29 million tons by 2040. The consequences are far-reaching. Plastic waste endangers marine life, damages marine ecosystems and harms fisheries. Even though plastic is a valuable, low cost and versatile material, its overuse and mismanagement are clearly harming the planet and its oceans in particular. Tackling marine plastic pollution is an urgent sustainability issue.

There is still much to learn from ocean life. It is a largely unknown resource base at this time, but it is known to have huge quantities of fossil fuels and other valuable resources under the ocean floor in several locations. There is much exploration still to be carried out. It will afford marine scientists and biologists much interesting research in the decades ahead and, in particular, the special or rare metals that can be found beneath the seabed.

The gases which make up the air that is breathed on Earth, has currently the right content of oxygen in it, ±20%, to live comfortably. No oxygen and even 100% oxygen make life impossible. It is a major source of study for the astronauts on the various countries' space programmes to see how the human body reacts to long-term exposure to space conditions. So far, concentration has been on the Moon but within the century, NASA is expected to turn greater attention to one of the other planets or one of its satellites.

REFERENCES

Bagge, M., Klemann, V., Steinberger, B., Latinović M., Thomas, M. (2021) Glacial-isostatic adjustment models using geodynamically constrained 3D earth structures. *Geochemistry, Geophysics, Geosystems*. 22(11):1–21.

Bawden, T. (2024) Our planet is shifting on its axis, *The i– Newspaper*, November 25.

Bennet, P. (2022) Global forest area declined by 60% since 1960. *EcoWatch* August 3, 2022 Available at: https://www.ecowatch.com/global-forest-decline.html

Cunliffe, B. (2012) Britain Emerges. Chapter 2. In *Britain Begins.* Oxford University Press, Oxford: 35–76.

IBM (2023) *What is a smart city*? Available at: https://www.ibm.com/think/topics/smart-city

Kirwan, C. G. (2024) *Smart Cities and Artificial Intelligence*. Elsevier, Holland.

Mann, T., Bender, M., Stocchi, P., Vacchi, M., Switzer, A., and Rovere, A. (2019) Holocene sea levels in Southeast Asia, Maldives, India and Sri Lanka. *Quaternary Science Reviews* 219(1):112–125.

Naess, P. (2001) Urban planning and sustainable development. *European Planning Studies* 9(4): 503–524.

Ritchie, H. (2024) *Not the End of the World*. Chatto & Windus, London.

Scindo (2024) *Creating Innovative Enzyme Systems for a Sustainable Future*. Available at: https://scindo.bio/

Searchinger, T., Waite, R., Hanson, C., Ranganathan, J., and Dumas, P. (2019) *Creating a Sustainable Food Future: A Menu of Solutions to Feed Nearly 10 Billion People by 2050.* Final Report. World Resources Report.

UN (1987) *Our Common Future (The Brundtland Report)*. Report of the World Commission on Environment and Development. Available at: https://sustainabledevelopment.un.org/content/documents/5987our-common-future.pdf

UN (2024) *World Population Prospects 2024*. Department of Economic and Social Affairs Population Division. Available at: https://population.un.org/wpp/graphs?loc=900&type=Demographic%20Profiles&category=Line%20Charts

Varma, P. (2019) Adoption and the impact of system of intensification of rice yields on household income and an analysis for India. *Journal of Applied Economics* 52(45):4946–4972.

Wikipedia (2025) *Clean Air Act 1956*. Available at: https://en.wikipedia.org/wiki/Clean_Air_Act_1956

World Bank (2025) *How the World Bank Group is Addressing Marine Plastic Pollution*. Available at: https://www.worldbank.org/en/topic/how-the-world-bank-group-is-addressing-marine-plastic-pollution

WWF, (2025) *Soil Erosion and Degradation*. World Wild Life Fund, Washington, DC. Available at: https://www.worldwildlife.org/our-work/forests/soil-erosion-and-degradation/

Younger, J. S., Parry, D. E, Lubis, H. A., MacLernon, A., Wignall, D. J., Hasan, D., and Benton, G. G. (2015) Greater Jakarta: The world's second largest conurbation. *Proceedings of the Institution of Civil Engineers – Municipal Engineer* 168(4):253–261.

Energy

6

David Boyd, Scott Younger, and David Meigh

6.1 INTRODUCTION

Energy has been at the core of our development as a species ever since our ancient ancestors recognised the advantages of fire for essentials such as cooking, defence and weapons production. Initially wood was the principal external source of this energy, gradually joined by water and wind as well as other fossil fuels such as peat, coal, oil and gas. This dependency on energy, and particularly fossil fuel, grew exponentially as most developed nations moved at varying pace towards industrialisation during the late 18th century in Europe and early 19th in USA.

The downsides of energy production in terms of pollution gradually became apparent during the early decades of the Industrial Revolution, as smoke and effluents became more obviously a danger to health and safety. Only much later in the late 20th century did the longer term risks from greenhouse gases (GHGs) become linked with fossil fuel-based development and permanent changes to our atmosphere and climate.

This chapter outlines the dependency of mankind's historical development on various fossil fuels and then focusses on the example of nuclear energy as a crucial long-term alternative to the continued dependency on these fossil fuels. Other options such as wind, hydro and solar power are briefly covered as options, but the core message is the role that nuclear energy has to play in our future sustainable development.

DOI: 10.1201/9781003566274-6

6.2 FOSSIL FUELS

6.2.1 Historical Context

Fossil fuels such wood, coal, oil and gas have been the main sources of energy and the key drivers of development in Europe for just about the last three centuries; in USA for the last two centuries and in China for less than 100 years. Initially wood, then later coal, were the principal energy sources, and only since the beginning of the 20th century have oil and gas become increasingly important. Initially, it was principally coal until usage of oil gradually increased as the 20th century advanced, especially from 1920 to 1930s for transport. However, while use of coal steadily declined in Europe from the 1970s, this has been more than offset by China's industrial expansion in the last 40 years, which was significantly based on coal.

The use of coal as a major energy source has a varied history. In the UK, by the early 1970s, there was recognition that the coal industry was in decline. At the same time, the Organization of the Petroleum Exporting Countries (OPEC), mainly the Middle East oil-producing countries, in 1973 put up the oil price by a factor of 4. Coincidently around this time North Sea oil became available and this combined with the rise in crude oil signified the beginning of the end of the coal industry in UK, although it did lead to more than a decade of some industrial unrest.

In contrast, China's coal use has more than doubled since the turn of the century, and it is reputed that China puts a new coal-fired power station on stream every month. China accounts for more than half of the 2,130 GW global capacity of coal usage. The USA and India are the next largest users, about one-fifth that of China with Japan, South Africa and Indonesia coming next. Indonesia is a main supplier of coal to China, Vietnam and India.

Some states of the USA, while investing in renewables, are still heavily dependent on coal and gas-fired power stations for their prime sources of energy. There is a vocal sizable minority that does not believe in climate change, which will become an increasing political issue in US government circles in the times ahead.

In the latter stages of the 20th century, increasing alarm was felt because the climate appeared to be changing in different parts of the world. Long-term changes in world climate had long been recognised on a geological timescale and manifested through a series of Ice Ages; however, 'global warming' was now increasingly linked to fossil fuel-induced CO_2 emissions increasing and escaping to the atmosphere as reported by the IPCC in 1990 (IPCC, 2010). There is now a UN-led mandate for all countries to reduce their carbon emissions to net zero, preferably by 2050. Some, led by China and India, say that target date is too soon for their planned development, and 2070 is a more reasonable target.

There is an increasing emphasis on the renewable energy, which is derived from natural sources that are replenished at a higher rate than they are consumed. These sustainable sources include, such as solar, wind, hydro, marine (tidal and wave) movement, and geothermal heat. It is seen that this renewable energy, or clean energy by way of nuclear, can largely replace fossil fuels by the end of the 21st century.

6.2.2 Coal and the Changes It Brought About

Coal has been formed by the compaction or lithification of ancient trees, ferns and other plant debris over considerable periods of time. In the UK, the principal coal deposits come from the Carboniferous era (300–360 million years ago); in contrast, peat deposits in Ireland and East Anglia were formed within last 10,000 years. The United States Geological Survey (USGS, 2024) divides coal into a series based on increasing compaction/lithification and carbon content and decreasing volatile content: peat (the unlithified form), lignite, bituminous coal and anthracite (the lithified forms).

Peat and near-surface or outcropping coal have been used as sources of heating, cooking and smelting for thousands of years. In the UK, coal is known to have been used by the Romans extracted by near surface or shallow single-shaft mining techniques. Coal continued to be used in UK through the Middle Ages, and mining is reported in 1575 in the Lowland belt of Scotland, but its use really took off towards the middle of the 18th, beginning of the 19th century. It was first used in the USA in 1701, in India from 1774 and in Australia at the end of the 18th century. Its value as an efficient resource for heating water for steam-driven pistons, opened up a whole range of industrial opportunities and its commercial use really burgeoned. This was an early pointer to the start of the Industrial Age, although canals for transport of bulk goods were put in place a few years earlier than this. The stirrings of the Age were overall in the last half of the 18th century.

The coal-fired steam engine was largely unchallenged for many decades in terms of efficiency and the enormous power it provided for industrial purposes. Its pre-eminence lasted until the early decades of the 20th century when oil and its derivative petroleum increasingly offered a better alternative for some purposes, e.g. for driving the motor vehicle, and increasingly this grabbed the headlines. This followed from Scottish inventor, Manchester-based chemist James Young's experiments deriving paraffin from coal (1847). In addition, oil, which was discovered from the mid-1860s onwards in significant deposits, came into increasing prominence as the 20th century unfolded.

The story of coal, as said, dominated industry for about a century and a half. The Industrial Age through the 19th century and first half of the 20th could just as well be called the Age of Coal (Paxman, 2022). It changed industry significantly, allowing many new ones to open up, and people in all walks of life were affected, whether directly or indirectly.

Large-scale coal mining was undertaken in Britain in the North of England area, but this soon spread to other places where the geology indicated the Coal Measure deposits occurred in viable quantities, in England, in Scotland's central belt and in South Wales. The last deep mine in UK closed in 2015 and the last opencast mine in 2023. In Western Europe, coal is still mined in several countries, particularly Poland and Germany. Across the world, China, Indonesia, Australia, South Africa, India and Vietnam still undertake significant deep and near-surface mining.

In Indonesia, the coal mining industry has progressed over the last few decades from a very small domestic coal supplier to the fifth largest coal producer in the world, mining over 430 million tonnes in 2016 due to the advent of large-scale opencast mining operations (Younger et al., 2018). Indonesian coal is sourced both for domestic consumption and as an export product. A recent development is the re-activation of coal mining in some of the traditional areas in the Far East, influenced by the growing need for electricity, economics of importing fuel and difficulties in the nuclear power industry.

In Australia, the high-quality, black bituminous or coking coal, is found in Newcastle, New South Wales and Queensland, whereas poorer quality, high-water content lignite is mined in Victoria and South Australia. Australia exports the majority of its coal, especially the coking coal, to the countries of East Asia – Japan, South Korea, Taiwan, China – and further, to India. India is a net importer of coal, despite producing a significant amount of what it needs, but its coal contains a high ash content.

6.2.3 UK Coal Mining

In the beginning in Britain, the owner of the land in which the coal was found, usually at depth, had sole rights to its exploitation, and this held until the coal industry was nationalised in 1948. The early mines had only a single shaft until owners found they really needed two shafts to circulate the air and make the operation safer underground for the miners. Paxman (2022) opens his account of the coal industry with a dreadful mining disaster in 1862, with an operation which had only one shaft where a main beam suddenly broke, effectively closing off the shaft and use of the mine. Apart from the men killed directly, the accident left over 150 miners, nearing the end of their shift, stranded. All died, because there was no access to provide rescue.

In the first days of coal mining, there was no legislation preventing anyone to go down a shaft, and women were also employed with tasks like hauling, as well as acting as pit 'ponies', and children as young as five to get to tight places that the adults could not reach. It was only in 1842 that the Westminster parliament passed a bill preventing this practice. The Mines and Collieries Act of August 1842 precluded women and girls and boys under aged 10 from working underground. This achieved much popular acclaim, although The House of Lords, which included many landowners, opposed much of the legislation and dragged their feet until the law was enacted.

At its height, 224 million tons of coal were mined on average annually, and Paxman (2022) knowledgeably describes how coal mining helped to bring about the change to society that altered British life forever, and communities in Europe and North America quickly followed (Figure 6.1, Ritchie, 2019).

It was a filthy and dangerous occupation with rockfalls, explosions from fire damp and water leakages from faults in the rocks, and this realisation bred a toughness in the men working in the industry. In the late 19th century there was several localised miners' trade unions, the banding together of workers to stand up for their human rights. The power remained in the hands of the landowners until 1888 when a national trade union was formed in Newport, South Wales. "By 1910, the Miners' Federation of Great Britain had almost 600,000 members and was the biggest trade union in the country" (Paxman, 2022).

At the beginning of WWI, there were over 1 million miners working in Britain's coal mines. Labour politicians, especially those close to the industry, were decidedly 'cool' about participating in the war since it was the opposite of the doctrine of international fraternity, and they made their voices heard, so nurturing the emergence of socialism. This reaction from the Miners' Federation marked the growing power of the

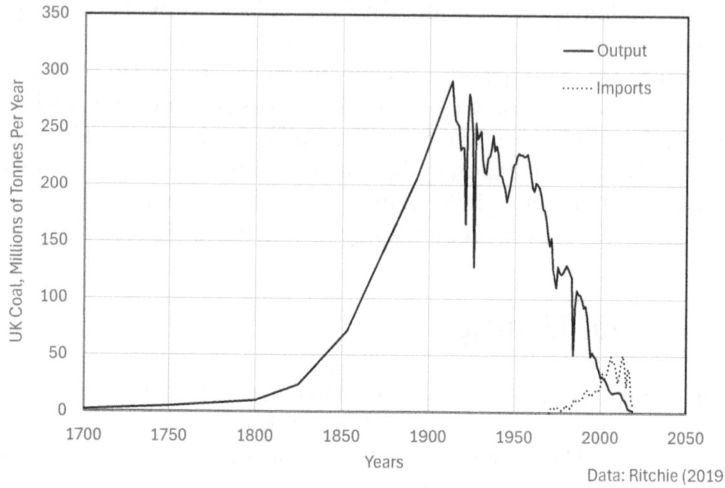

FIGURE 6.1 UK coal production and import. (Data: Richie, 2019).

industry and its strong socialist roots it pervaded. This was particularly notable at the start of World War II (WWII) when the Soviet Union, led by Stalin, was on the Nazi side; some miners refused to participate until the Soviet Union was forced to change sides when Hitler's Germany invaded Russia in August 1940.

After the war was over and a new Labour government was elected in 1946 under Clement Atlee, the coal mining industry was nationalised, among other national assets, and henceforward called the National Coal Board. Coal mines were no longer owned by private landholders.

The beginning of the end of the British coal mining industry took place in the late 1960s. The miners were indulged by a Labour government, while unprofitable mines were closed down and cheaper coal was obtained from abroad; workers left the industry steadily for pastures new. The industry was dying just as the new game in town, oil, was discovered offshore in the North Sea in commercial quantities and was really taking off, in a financial environment that was becoming favourable for oil, since OPEC put up the price of oil by a factor of four in 1973.

The 1970s were a turbulent time for British politics and the economy, with four prime ministers in a decade – Heath, Wilson, Callaghan and Thatcher. In addition, the UK's economy depreciated considerably in 1977 when Labour's Jim Callaghan had to rely on the International Monetary Fund to bail it out. The miners' trade union was becoming 'difficult', and unreasonable to deal with, except in the Labour government years. Miners feared they were in a dying industry, and they finally had to accept this in 1981, when Thatcher was prime minister. Her policies discouraged further mining of coal. Coal, however, was still mined in several countries across the world, e.g. the USA, China, India, Australia, Indonesia, Poland, Russia and South Africa to name some of the most prominent.

Noting today's worldwide discouragement of use of fossil fuels by the UN, because of its significant contribution to CO_2 in the atmosphere, threatening global warming, some countries have cut out the use of coal in their energy mix, while others have delayed a cut back, because its use is still key to their economic development. Most notable in pursuing an energy policy that uses coal in the mix are China and more modestly India, the two countries with the largest populations, some 36% of the total world population, and they will be significantly relying on it for the next few decades.

China greatly expanded its use of coal from 2003 till 2016. In 2023, they used 4 billion tons, which was more than half the world's total produced. New coal-fired power stations are still being built. China is taking a significant quantity of coal from Australia and Indonesia to top up their own sources, inadequate for all their requirements. Ships are also supplying coal to India and Vietnam from Indonesia. Indonesia is also using their coal for domestic purposes. Their deposits are almost exclusively at or near the surface and thereby suitable for opencast methods of mining. Characteristically the Indonesian coal is geologically young and comprises predominantly lower-ranked material such as sub-bituminous and lignite ('brown') coal, primarily only of use for electricity generation rather than steel production, for example.

6.2.4 Coal and Energy

Historically coal has been associated with two principal energy production processes; coal gas (sometimes referred to as 'town gas') and coal-fired power stations.

In the late 19th and early to mid-20th centuries, prior to the development of natural gas, almost all gas for fuel and lighting was produced by the distillation of bituminous coal. The gas was driven off by heating the coal in air-tight chambers, storing it in 'gasometers', and distributing it through a network of pipes to houses and other buildings where it was used initially for lighting and, later, for cooking, heating and some industrial processes.

The first coal-fired power plant in UK came on stream in 1882 at Holborn Viaduct, designed by Thomas Eddison. Also in 1882, in New York, the USA opened their first coal plant. By the end of WWII, 97% of the total energy supplied in UK came from coal. However, the last remaining coal- fired plant is due to close in 2024, assuming that there is enough renewable, wind or solar, or nuclear capacity built to replace it, thus bringing a century and a half of coal-fired power to an end in Britain. However, as there is research taking place to remove CO_2 from the atmosphere, there may still be a place for coal in the energy mix of the future.

In the USA, there is also a plan to rely less on coal-fired power, and it has shrunk to only supplying approximately 20% of the country's energy needs. Against this, in the past two decades the Chinese, as noted previously, have built many coal-fired stations, such that they have over 1,200 of them and produce more than 50% of the world's total coal-fired output for power. In fact, East Asia plus India and China, together produce some 75% of the world's supply of coal-fired energy. Germany, like Britain, formerly a big producer of coal-fired power for its heavy industries, has reduced significantly its reliance on coal and Europe's use generally is declining annually. China has formally said that coal would be a major part of its energy mix for the next 50 years. However, it has also built and expanded its renewable resources, mainly solar, albeit this is not expected to be a major source of energy in China for the whole of the 21st century.

Coal in the late 19th century up to the mid-20th century had a crucial role in providing energy for domestic use in most European countries. In addition to coal itself, domestic gas was produced by the destructive distillation of bituminous coal (so-called 'town gas'). The gas was driven off by heating the coal in air-tight chambers, storing it in 'gasometers', and distributing it through a network of pipes to houses and other buildings where it was used for lighting, cooking and heating.

Looking at the rest of this century, it would seem that, while use of coal to provide the energy that is needed will decline, especially after the next two or three decades, coal usage will not make up other than a very minor portion of the world energy requirement from the next century onwards.

6.2.5 Oil

There had been many references to the presence of oil from about 4000 BCE by Herodotus, by the Babylonians in 1200 BCE and periodically since then. The production of the oil was almost exclusively from surface seeps. Its proper value was not realised until the 19th century, particularly the second half, when it was seen as an alternative to coal, e.g. as a means of supplying traction to lighter engines and for making it later into other useful organic-based by-products.

The internal combustion engine by which many millions of motor vehicles, heavy goods vehicles and ships have been powered for more than a century is the major user of oil products, petroleum and diesel, which came on stream in the 1920s.

Oil is expected to be the main source of fuel, as petrol and diesel, for the next two decades, although there is a growing market for hybrid vehicles, whereby the vehicle moves under electric battery from starting up and low speeds. This will reduce the problem of big city pollution in which a traffic jam means hundreds of motor vehicles standing, engines idling and polluting the atmosphere with dangerous carbon monoxide. The hybrid vehicles would seem to be a good solution for the next few decades until the all-electric vehicle is fully developed, assuming there is enough lithium to be found and mined to produce the enormous number of batteries required. Perhaps a replacement may be found? However, there is a growing desire to reduce the carbon emissions from exhaust fumes, particularly in the centre of large cities, and there are now the means available.

There are a few all-electric cars on the roads as of 2024, but they are relatively expensive, and there are not enough charging stations to make this a viable option for owners carrying out many long journeys, with one or two exceptions. However, some countries have been actively preparing to take electric vehicles. For example, in 2022, in Norway 80% of sales were of electric vehicles, in Sweden 32%, Netherlands 24%, and China 22% with 4 million sales.

Perhaps carbon emissions will reduce in two or three decades. An all-electric transport mode may not be necessary, for instance due to the provision of the liquid primary energy being sufficiently cleaner, and the call on fossil fuels being reduced. Is this in the near or not too distant future?

6.2.6 Natural Gas

Natural gas, which is obtained from deposits or from the offshoot of refining oil, is used widely and directly for heating, cooking, in electricity power stations and in chemical production.

There are 11 power stations in England and one in Scotland currently operating, but the power generated from gas-fired stations is waning and being superseded by offshore wind mainly. In 2016, Britain produced 119 TWh of gas energy, but in 5 years, this had dropped to 95 TWh, and the aim is to reduce this further. The USA has 1,900 natural gas-fired stations, much more than any other nation, and this produces 38% of their energy offtake. They have indicated that they will replace this gradually with renewable energy although it is subject to some political debate. China is currently investing in gas and nuclear, but their usage is much less than coal. It is more likely to take on renewable, e.g. solar solutions, and continue with nuclear for its future power requirement once it stops building coal-fired power plants.

6.2.7 Fossil Fuel Overview and Summary

In general, over the rest of this century, there will continue to be much use made of fossil fuels, albeit on a declining basis. However, it can be seen, and in the pages that follow, there is plenty of energy to supply the world's needs for the foreseeable future, this century and beyond. The challenge is how quickly the transition can be made to the non-fossil fuel types of energy to reduce the carbon emissions in the atmosphere. There is a long way to go. Data produced by the International Energy Agency (IEA) showed the minimal reduction in fossil fuel energy in the decades prior to the COVID epidemic (Table 6.1).

Decreasing use of fossil fuels will be the aim for the short to intermediate times; in reality, this may take 50 years because of the perceived industrial needs of China and reluctance of some countries including the USA, which is heavily invested in coal- and

TABLE 6.1 World primary energy supply by fuel percentages (IEA, 2021)

	PERCENTAGE ENERGY BY FUEL	
SOURCE	*1971*	*2019*
Coal	20.7	26.8
Oil	44.0	30.9
Natural gas	16.2	23.2
Nuclear	0.5	5.0
Hydro	1.9	2.5
Biofuels	10.8	9.4
Other renewables	0.2	2.2

gas-fired power stations, to embrace renewables. However, in the long term, fossil fuels are not expected to provide other than a minor part of energy requirements, except for specific industrial purposes. At the moment, the major oil companies are meeting the worldwide demand, and most are not doing much to transfer investment to renewables, with few exceptions. The major companies operate internationally and are very large and not generally answerable to state governments. Some are more sympathetic to climate change than others and are making long-term adjustments to accommodate increasing involvement in renewables. There is also a large number of sizeable companies that are state-controlled and are answerable to their respective governments. Not all take on the message of climate change; it depends on their seriousness towards this issue.

If there is to be a significant reduction in the output of carbon emissions, there has to be a considerable reduction in use of coal in China, and gas and coal in the USA, which they are supposedly trying to do. Britain is taking its reduction in use of fossil fuels quite seriously but its overall impact in world terms is small. It is assessed that Britain is responsible for about 1% of total world carbon emissions. However, if the reputed current level of reduction in carbon emissions across the world can be maintained or even further reduced, then we should witness within the next five decades reaching 0°C. Some of the nations of the world are following the COP21 (2015) mandate and taking fossil fuel reduction seriously, but not all, and more needs to be done.

It should be remembered, however, that before 1800 the population of the world had not reached 1 billion and the CO_2 emissions were 315 ppm. In 1960, the population had gradually increased to 3 billion with CO_2 emissions recorded as 335 ppm, i.e. a small factor over 160 years with a gradual increase in population of 2 billion and a significant increase in the use of coal. The current large increase in population to over 8 billion in just 60 years and the fast rise in carbon emissions to 420 ppm would seem to equate and climate changes to result. Carbon emissions indicate other human-induced factors than solely due to power generation as discussed later, e.g. the environment.

The UN states that the world has to reverse the trend of rising carbon emissions using fossil fuels, even allowing for the increase in population, hence the call to reduce usage of fossil fuels. The past records would suggest, however, that the climate can handle a reasonable level of CO_2 emissions, up to about 350 ppm. With a slow-down in the population growth, as forecast for the end of the century and countries taking action to reduce their use of fossil fuels, it ought to show a lower, improved level of carbon emissions. Similarly, the planned improvements to the environment regarding pollution should contribute to the reduction in carbon emissions.

6.3 NUCLEAR ENERGY

6.3.1 Public Acceptance

If there is one topic that is certain to cause a polarisation of views in the general population, then that topic is nuclear power. Is this because of ignorance of the technologies involved, fear associated with nuclear accidents, nuclear weapons or radioactive wastes, or an accumulation of social pressures against nuclear power?

People who work in the nuclear industry or live alongside nuclear facilities tend to be quite blasé about these concerns and try to explain the benefits of nuclear power as over-riding its dangers, and they stress the overall excellent safety record over its nearly 70-year history, Chernobyl and Fukushima notwithstanding. And indeed, they are mostly correct. Nuclear is a relatively safe and highly beneficial industry, especially compared to fossil fuel alternatives. But the concerns persist and cannot be ignored.

6.3.2 Fission versus Fusion

There are two very different nuclear energy producing options that need to be understood and differentiated in terms of process, cost, risk and return before being able to appreciate their role in future sustainable power generation. They are nuclear fission and nuclear fusion.

Fission is the process that powers nuclear plants today and involves splitting large atoms like uranium and plutonium into smaller atoms like caesium and strontium (and creating small amounts of exotic large atoms in the process, actinides like curium, americium and neptunium). These fission products can be highly radioactive and will remain so for a very long time.

In the fission reactor, the process involves the excitation of the nucleus of a fissile isotope by external neutrons causing fission (splitting the atom) releasing further neutrons (and large amounts of energy), which induce further fission in nearby fissile material and release more neutrons, causing a chain reaction (criticality). This may include a 'starter' neutron source that ensures that a chain reaction will begin.

The reaction requires sufficient mass of fissile material to be concentrated in close proximity (as in the core of the reactor) for the criticality to develop. The process is controlled or shut down by introducing neutron absorbing materials (such as boron rods) into the reactor, via a system of control rods which prevent the criticality continuing. Rapid shut down is achieved by a SCRAM system, which forces control rods into the core and is triggered automatically if system problems are detected, or in adverse environmental circumstances (such as earthquake), or it may be triggered manually.

Fusion is the process of causing small atoms (isotopes of hydrogen) to join together to form larger atoms (helium). This is what happens in the sun. Fusion produces only very small amounts of radioactive waste (some irradiated materials), but is still at the research stage, after almost 40 years, with several types of small-scale experimental reactors in operation. However, the world's first full-scale fusion reactor is being built at Cadarache in France (the International Thermonuclear Experimental Reactor – ITER, 2025), by an international consortium of countries, and a follow-on plant (DEMO) is under consideration as a full-scale power generation demonstration plant.

There are also other emerging fusion technologies being considered as capable of being scaled up to future power reactors, such as a smaller footprint spherical tokamak[1] – STEP (UKAEA, 2024) being proposed for a site at West Burton in Nottinghamshire and a magnetised target fusion (MTF) machine (General Fusion, 2025) being proposed at Culham in Oxfordshire, where the UK's centre for fusion research has been based since 1965.

There is also the National Ignition Facility (NIF) at Lawrence Livermore Laboratory in the USA (LLNL, 2025; Martin, 2012) that has been undertaking fusion research for many years using laser technologies. So, fusion is moving forward, but much scientific research and technological development are needed before an operational fusion power station becomes a reality. Fusion thus remains the great hope for a sustainable energy future.

6.3.3 Low-Carbon Energy Knowledge

The current nuclear industry (fission) considers itself a low-carbon technology that can contribute significantly to the global decarbonisation agenda and provide energy security for today's generations. It recognises the concerns regarding safety, proliferation and waste management and has active programmes and systems in place that have been addressing these over very many years. The technology base is large and diverse, but this can seem impenetrable to those not scientifically trained, which can be a problem in trying to build public support for nuclear power going forward. Understanding the range and complexity of the nuclear industry is not something you can pick up easily.

Institutions like the International Atomic Energy Authority (IAEA), established under the auspices of the UN, and industry bodies such as the World Nuclear Association (WNA) and national associations such as the Nuclear Industry Association in the UK, all seek to provide information and guidance on nuclear matters and seek to dispel exaggerated claims of danger to the public. Their websites provide huge amounts of information with significant transparency. Much of the information presented in this chapter has been reproduced or paraphrased from websites by these organisations.

For instance, the IAEA (2024) list some 413 working nuclear power reactors across the world in December 2023 and a further 192 reactors permanently shut down, in 34 countries. This totals some 18,772 reactor-years of experience. A further 69 reactors are known to be planned, not counting small modular reactors (SMRs) which are gaining support. So, there are lots of people out there with high levels of knowledge and experience of running nuclear reactors safely, and the industry is generally considered strong at present. That is not to say that we should not be concerned or be complacent. Indeed, fears of uncontrolled events, nuclear proliferation and long-term safety of nuclear waste are real and require constant vigilance at both the technical and political levels, particular in zones of conflict.

6.3.4 Nuclear Energy and Sustainability

So, if concerns about fission are real and fusion is still years away, is nuclear sustainable? The Brundtland definition of sustainability is "meeting the needs of the present without compromising the ability of future generations to meet their own needs". This is particularly problematic for nuclear energy. Large nuclear plants take a long time to plan, gain regulatory approval and build, but they run for a long time. For instance, Hinkley Point B in the UK ran for 46 years from 1976 to 2022. In that period, it generated 311 TWh of electricity, enough to meet the needs of every home in the UK for almost 3 years, so a major contributor to meeting energy demand (EDF, 2025a).

Plants also take years to defuel and decommission, and then there is the issue of long-term management of nuclear wastes. So, nuclear plants in operation today and planned plants for the future, while meeting the needs of today's generations, but will inevitably carry over to impact on future generations, and regarding nuclear wastes will impact for very many generations to come. Thus, it is incumbent on today's generation to take action to mitigate the potential dangers to future generations and particularly the management of long-lived radioactive wastes.

6.3.5 The Nuclear Wastes Issue

Nuclear power is characterised by the very large amount of energy produced from a small amount of fuel and over a small footprint. The amount of waste produced during this process is relatively small, but much of this is dangerously radioactive and therefore must be managed carefully as hazardous material. A later section of this chapter looks at the waste issue in more detail including technical solutions. The costs of managing and disposing of waste and the costs of power plant decommissioning are usually included as part of the electricity cost to consumers and may be accumulated as sequestered funds or government guarantees, and the financial burden on future generations may thus be mitigated.

Forward provisions like this require political and economic stability to be maintained through to disposal. Thus, the waste issue becomes both technical and political, and because the nuclear industry is so diverse across the world, this requires international cooperation with bodies such as the IAEA being vital to success (IAEA, 2021).

In 2022, the IAEA quoted Mikhail Chudakov, IAEA Deputy Director General and Head of the Department of Nuclear Energy as saying "The successful management of radioactive waste and spent nuclear fuel (SNF) underpins the peaceful use of nuclear technology in all its facets, from low-carbon energy production to medicine and industry". A major conclusion is that effective and long-term management of wastes must be achieved for nuclear to be seen as part of a sustainable energy future.

6.3.6 Summary Figures

The WNA (2025) reports that the figures in Table 6.2 may be regarded as typical for the annual operation of a 1,000 MWe nuclear power reactor using 4.5% enriched fuel and with 45 GWe/t burn-up.

This summary allows the large scale of the nuclear fuel cycle to be appreciated, especially if the equivalents of 500 such power stations are to be operating by 2030.

6.3.7 Nuclear Generation

Most nuclear plants in operation today, and currently planned, are large plants delivering steady, reliable, base-line power. Typically, each reactor will have a capacity of between

TABLE 6.2 Typical annual data for 1,000 MWe nuclear power reactor

OPERATION	*ANNUAL REQUIREMENT/OUTPUT*
Mining	Between 20,000 and 400,000 tonnes of uranium ore
Milling	192 tonnes of uranium oxide concentrate (which contains 163 tonnes of uranium)
Conversion	241 tonnes of uranium hexafluoride, UF^6 (with 163 t U)
Enrichment	25 tonnes of enriched UF_6 (containing 16.85 t enriched U) – balance is 'tails' @ 0.22%
Fuel fabrication	19.1 tonnes UO_2 (with 16.85 t enriched U)
Reactor operation	8,760 million kWh (8.76 TWh) of electricity at 100% output, hence 18.6 tonnes of natural U per TWh
Used fuel	19.1 tonnes containing 200 kg transuranics (mainly plutonium), 18.3 t uranium oxide (<1.0% U-235), <0.6 t fission products

Source: WNA (2025).

600 and 1,600 MW, and some sites will have several reactors such as at Hinkley Point C in the UK, which will have twin European Pressurised reactors (PWRs) each capable of generating 1,630 MW for 60 years (EDF, 2025b).

By 2030 there may be close to 500 large reactors operating across the world. The most prominent reactor type is the Pressurised Light-Water Moderated and Cooled Reactor, sometimes termed PWRs; of which 302 were operational in 2020, followed by the Boiling Light-Water Cooled and Moderated Reactor (termed BWRs) of which 83 were operational in 2020 and then other reactor types including the Pressurised Heavy-Water Moderated and Cooled Reactor (PHWR, 48 operational in 2020). Recently, SMRs are gaining prominence, typically below 300 MW capacity but some slightly larger models are proposed including the 470 MW by Rolls Royce in the UK. These would be suitable for stand-alone networks or supplying a small city or linked to hydrogen production. Some very small reactors as low as 10 MW are proposed for remote locations and indeed such reactors have been in use for several years in remote areas in the Russian Arctic.

All these reactors have a core of fissile material, which is atomic material capable of splitting into smaller atoms and so releasing the energy holding them together. This energy is emitted as heat (and radioactivity). The heat is collected by fluid, usually water (or carbon dioxide in the old Magnox and Advanced Gas-Cooled Reactor (AGR) reactors in the UK) circulating in a closed loop through the core which is then used to generate electricity using the same principles as in fossil fuel power stations, generally via steam turbines. Some specialised or prototype reactors use helium as the circulating fluid which may drive a gas turbine. In effect, the reactor core is a hot box akin to a coal boiler. This is used to produce steam to drive steam generators.

In the BWR, the water circulating through the core is turned directly to steam driving the steam generators and is then condensed and returned to the core creating a closed loop system. In the PWR and PHWR, the water circulating through the core passes through a steam generator vessel where heat is exchanged across a physical interface to a secondary closed-loop water system which generates steam to drive the steam turbine. The steam is then condensed and returned via the closed loop.

At the end of this arrangement, it is necessary to remove heat from the system to condense the steam. This is referred to as the heat sink and is typically a further water circulation system either river or seawater, taken in via a water intake and returned via a water outfall, where the intake and outfall are separate and do not interfere with each other (upstream intake and downstream outfall for a river or physically distanced intake and outfall in the sea). Alternatively, the heat sink may be a further closed loop system (with make-up water) using cooling towers or exchange to air via cooling stacks. Thus, unlike the primary coolant, the heat-sink water has never been in contact with the core and so is not susceptible to radioactive contamination.

The reactor and steam generation system is referred to as the Nuclear Steam Supply System (NSSS), various forms of which (PWR, BWR, PHWR, etc.) have been developed by different suppliers. Efforts have been made by suppliers to standardise their form of NSSS both to aid in approval and regulation and to have a fleet of similar reactors which can benefit from mutual experience, and a shared supply chain. However, all sites are different and so the foundations for structures and balance of plant arrangements typically have to be adjusted to suit site conditions.

Radioactivity is controlled by containment and shielding structures and by using remote handling equipment for accessing and moving radioactive components. Regular monitoring of exposure to radioactivity is intrinsic to operations and includes fixed systems with alarms and personal dose meters worn by operatives. This is coupled to regular health checks on staff which also pick up other non-industrial health issues, which is beneficial to overall health, and in the UK, explains in part why the nuclear industry shows a better Health and Safety record than other comparable industries.

6.3.8 Nuclear Waste Management

Radioactive waste is not unique to nuclear power generation. Radioactive materials are used in medicine, agriculture, research, manufacturing, non-destructive testing and minerals exploration. All these materials need to be managed and disposed of safely.

Nuclear waste may be categorised as low, medium or high level based on the amount of radiation that it emits:

1. Low-level waste (LLW) has a radioactive content not exceeding 4 gigabecquerels per tonne (GBq/t) of alpha activity or 12 GBq/t beta-gamma activity. LLW does not require shielding during handling and transport and is suitable for disposal in near-surface facilities.
2. LLW is generated from hospitals and industry, as well as the nuclear fuel cycle. It comprises paper, rags, tools, clothing, filters, etc., which contain small amounts of mostly short-lived radioactivity.
3. LLW comprises some 90% of the volume but only 1% of the radioactivity of all radioactive waste.

Intermediate-level waste (ILW) is more radioactive than LLW, but the heat it generates (<2 kW/m^3) is generally not sufficient to require cooling in the design or selection of storage and disposal facilities. However, due to its higher levels of radioactivity, ILW requires shielding.

ILW typically comprises contaminated filters, resins, chemical sludges and metal fuel cladding, as well as contaminated materials from reactor decommissioning. Large items may be size-reduced and along with smaller items and any non-solids may be solidified in concrete or bitumen and packaged for disposal. It makes up some 7% of the volume and has 4% of the radioactivity of all radioactive waste.

High-level waste (HLW) is sufficiently radioactive for its decay heat (>2 kW/m^3) to increase its temperature, and the temperature of its surroundings, significantly. As a result, HLW requires both cooling (heat dissipation) and shielding.

HLW arises from the 'burning-up' of uranium fuel in a nuclear reactor and contains the fission products and transuranic elements generated in the reactor core. HLW accounts for just 3% of the volume, but 95% of the total radioactivity of produced waste. There are two distinct kinds of HLW:

1. Used fuel that has been designated as waste
2. Separated waste from reprocessing of used fuel

HLW has both long-lived and short-lived components, depending on the length of time it will take for the radioactivity of particular radionuclides to decrease to levels that are considered non-hazardous. If the shorter-lived fission products can be separated from long-lived actinides by reprocessing, this becomes important in the management and disposal of HLW. HLW is the focus of significant attention regarding nuclear power, and is managed accordingly, as described subsequently.

It is important to recognise that we are surrounded by 'background' radiation all the time, which varies naturally from place to place and with variations in geology. Also, radioactivity is relatively easy to detect with modern equipment, and many everyday materials contain measurable radioactivity.

Nuclear waste which contains radioactive materials at a level which is not considered harmful to people, or the surrounding environment may be categorised as very low-level waste (VLLW). This consists mainly of demolished material (such as concrete, plaster, bricks, metal, valves and piping) produced during operation or decommissioning on nuclear sites. Other industries, such as food processing, chemical and steel, also produce VLLW resulting from natural radioactivity present in certain minerals used in the manufacturing processes. VLLW is typically disposed of to domestic landfill but may be placed in separate cells.

6.3.9 Waste Disposal

Studies have been undertaken over many years to determine the optimum approach to disposal of nuclear waste leading to a general consensus in favour of geological disposal.

Currently, there are no operational disposal facilities for SNF or HLW, but Finland's Olkiluoto repository is due to commence operations by the mid-2020s for disposal of SNF.

Some countries are more advanced than others in the move towards deep geological disposal. Plans for disposal of spent fuel are well advanced in Finland, as well as Sweden, France and the USA, though in the USA, there have been political delays (see Table 6.3). In Canada and the UK, deep disposal has been selected, and the site selection processes have commenced.

Total volume of SNF/HLW is relatively small, and the longer it is stored the easier it is to handle. Also, there is a reluctance to dispose of used fuel because it represents a significant energy resource which could be reprocessed at a later date to allow recycling of the uranium and plutonium. Thus, initially at least, SNF may be placed into a repository in a recoverable condition pending final decisions on its fate, and validation and verification of the repository design, before the repository is sealed permanently. There is no commercial value in HLW from reprocessing and eventually this will need permanent disposal.

A key requirement of deep disposal is that the sites must remain safe until radioactivity decays to a safe level, which in the case of actinides in reprocessed HLW and SNF may be over 100,000 years. Thus, the repository must be geologically stable over a long period and designed to prevent radioactivity escaping to the surface with a principal focus on preventing contamination transmission by groundwater flow.

Over such long time periods, other forms of transmission must be considered such as diffusion of radio nuclides under a chemical concentration gradient, thermal gradient or electrical gradient, as well as cross effects such as thermal gradients driving groundwater flows. ILW may be disposed of in a purpose-built repository or in the same repository as HLW but in cells designed differently.

6.3.10 Fusion Science

Fusion, the process by which lighter atoms are fused to make heavier atoms, is thought to be somewhat complicated and fusion science can seem impenetrable, but there is a huge compendium of information to peruse if one has the inclination.

Wikipedia provides an excellent and wide-ranging base level of information on the fusion process. In addition, the website Fusion Energy Base provides further information (Wikipedia, 2024).

Despite the complexity of the research and development, the science has progressed to the stage where practical fusion power appears a real possibility this century. The brief summary below gives an indication of the concepts involved.

As long ago as 1934, the first fusion was achieved using a particle accelerator to shoot deuterium nuclei into a metal foil containing deuterium. Since then, two basic approaches have been studied extensively and gained prominence, magnetic confinement fusion (MCF) and inertial confinement fusion (ICF). For fusion to occur, fuel material must be heated to tens of millions of degrees and compressed at high pressures, for sufficient time to allow collisions to take place. This combination of density, temperature and time is referred to as 'the triple point or triple product'.

Magnetic fusion works by heating a low-density plasma of deuterium and tritium (isotopes of hydrogen) at about 10^{14} ions per cm^3 (air is about 100,000 times denser) to

TABLE 6.3 Preferred waste options

OPTION	*SUITABLE WASTE TYPE*	*EXAMPLES*
Near-surface disposal at ground level, or in caverns below ground level (at depths of tens of meters)	LLW and short-lived ILW	• Implemented for LLW in many countries, including Czech Republic, Finland, France, Japan, Netherlands, Spain, Sweden, UK and USA. • Implemented in Finland and Sweden for LLW and short-lived ILW.
Deep geological disposal (at depths between 250 and 1,000 m for mined repositories, or 2,000–5,000 m for boreholes)	Long-lived ILW and HLW (including used fuel)	• Most countries have investigated deep geological disposal, and it is official policy in several countries. • Implemented in the USA* for defence-related transuranic waste at WIPP. • Yucca Mountain site identified in the USA**. • Preferred sites selected in France, Sweden and the USA. • Facility under construction and due to begin operations in 2023 in Finland. • Geological repository site selection process commenced in the UK and Canada. • Work on a Chinese deep repository has commenced for a site in the Gobi Desert.

Note: *In the USA, defence-related transuranic (TRU) waste – which has similar levels of radioactivity to some ILW – is disposed of in the Waste Isolation Pilot Plant (WIPP) deep geological repository in New Mexico. WIPP does not have a license for disposal of SNF or HLW.

Note: **In the USA, the Yucca Mountain site in Nevada had been chosen as a deep geologic repository for disposal of high-level radioactive waste, but soon after entering office, the Barack Obama administration decided to cancel the project. However, the Nuclear Regulatory Commission's Atomic Safety and Licensing Board (ASLB) noted that derailing the legislated decision-making process in this way may not be allowed, and the project remains in a state of limbo and controversy.

extremely high temperatures, around 20 keV (~200 million C) for the order of 1 second. A plasma is a heated (high energy) gas, where all the electrons have been stripped away and the gas has become fully ionised. The ions are positively charged, so they repel each other due to the electrostatic force. Fusion occurs when two ions collide at high energy, allowing the strong nuclear force to overcome the electrostatic force at a short distance, releasing energy.

Several experimental machines have been in existence for some decades to trial this method including donut-shaped vacuum vessels called tokamaks and spiral/helical-shaped donuts called stellarators. These machines use a strong magnetic field to confine the plasma. Heating is achieved by radio frequency/microwaves and by neutral beam directed into the plasma. So far these have been scaled-back machines that have been intended to investigate fusion science, with the amount of energy used to heat the plasma and provide the magnetic confinement to be greater than the energy released, although parity has almost been achieved at the JET facility at Culham in the UK.

The ITER tokamak design is a follow-on to JET and is currently being built at Caderache in France to test the magnetic approach at full scale and is expected to produce up to a ten-fold increase in output over input energy. DEMO will be EURO fusion's successor to ITER and will be a transition from a science-driven project to an industry-driven and technology-driven programme leading to production of electricity (ITER, 2025).

Inertial fusion occurs at much higher densities than MCF, at about 1025 ions/cm^3, about 100 times the density of lead, which causes the reactions to occur extremely quickly (~1 ns). The first ICF devices were the hydrogen bombs invented in the early 1950s. A hydrogen bomb consists of two bombs in a single case. The first, the primary stage, is a fission-powered device normally using plutonium, which surrounds the secondary stage, which is the fusion fuel (tritium and deuterium). When the first stage explodes, it causes the fuel inside to be driven inward, compressing and heating it to reach the temperature and density where fusion reactions occur.

The requirement for ignition by a fission bomb makes this method impractical for power generation. Consequently, research has focused on other methods to compress and heat the fuel. This is achieved by causing energy to be deposited on a fuel target (about the size of a pin head dosed with tritium and deuterium) using high-energy beams of photons, électrons or ions. Most ICF devices today use high power lasers. The so-called 'indirect drive', is where the beams shine not directly onto the fuel capsule but onto a small cylinder of heavy metal known as a 'hohlraum' which contains the fuel capsule. The beams are arranged so they the energy heats the hohlraum until it begins to give off X-rays which then fill the interior of the hohlraum and heat the fuel and cause the hohlraum to implode to fusion conditions.

Practical power plants built using ICF are known as inertial fusion energy plants. These would deliver several targets per second to the reaction chamber and capture the resulting heat and neutron radiation from their implosion and fusion to drive a conventional steam turbine. The most advanced research facility of this kind is the NIF in the USA, which started construction at Laurence Livermore National Laboratory in 1997, and in 2021 achieved 70% of the break-even point of energy out versus energy in (NLF, 2025).

In theory, the ICF approach could offer dramatically more gain than MCF because in MCF, the rate of energy loss to the environment is based on the temperature difference between the fuel and its surroundings, which continues to increase as the fuel temperature increases. In the ICF case, the entire hohlraum is filled with high-temperature radiation, limiting losses.

Other systems are under development including MTF, which is a concept that combines features of MCF and ICF. Like the MCF approach, the fusion fuel is confined at lower density by magnetic fields, while it is heated into a plasma which is injected into a liquid metal (lead) core created by rapidly rotating the liquid metal to create a central cavity into which the plasma is injected (General Fusion, 2025). As with the inertial approach, fusion is initiated by the injection process and by rapidly squeezing the target into a spherical shape via a precisely controlled compression system of synchronised pistons. This greatly increases plasma density and temperature to reach the triple product point and initiate fusion. The lead core also acts as a medium for heat transfer to a steam generating plant and hence power generation via a steam generator.

MTF aims for a plasma density of $10^{19} cm^{-3}$, intermediate between MCF ($10^{14} cm^{-3}$) and ICF ($10^{25} cm^{-3}$). At this density, confinement times are on the order of 1 μs. A demonstration plant is proposed to be built at Culham in the UK by a collaboration of General Fusion Inc. from Canada and UKAEA from the UK.

6.3.11 Nuclear as a Transition Technology; Comparison with Other Technologies

In the energy domain, many different units are quoted – joules, exajoules, million tonnes of oil equivalents, barrel equivalents, British thermal units, terawatt-hours, to name a few. This can be confusing and make comparisons difficult. Consistency can be obtained by converting all energy data to watt-hours which allows comparison of energy data across different metrics and sources.

Ritchie and Roser (2017) present a comparison of different forms of energy production and the consumption this would allow in different parts of the world currently. For instance, the NPP at Torness in Scotland is a 1,300 MWe capacity, gas cooled reactor (twin AGR type) commissioned in 1988 and scheduled for shut down in 2028, capable of providing electricity for almost 1 million people in the UK or about 10 million people in the developing world.

The contribution to electricity generation is shown in Table 6.4, extracted from the IEA report World Energy Outlook 2022 (International Energy Agency, 2022).

This relates to a scenario for reduced carbon emissions by 2050. It shows demand for electricity almost doubling in this period with renewables moving from 28% of the total to 65% and nuclear slightly falling from 10% to 9% of the whole, which is itself almost doubling in the period.

Table 6.4 shows nuclear power at 2,776 TWh in 2021, rising to 4,260 TWh by 2050, with installed capacity increasing from 413 GW in 2020 to 590 GW in 2050. If the WNA prediction of 500 operational large reactors by 2030 is correct, then the estimate of 590 GW of installed capacity in 2050 may be an underestimate. At the same time,

TABLE 6.4 World electricity sector

	STATED POLICIES SCENARIO (TWH)						SHARES (%)			CAGR (%) 2021 TO:	
	2010	*2020*	*2021*	*2030*	*2040*	*2050*	*2021*	*2030*	*2050*	*2030*	*2050*
Total generation	21,539	26,708	28,334	34,834	42,642	49,845	100	100	100	2.3	2.0
Renewables	4,234	7,539	8,060	15,073	24,442	35,452	29	43	65	7.2	4.9
Solar PV	32	824	1,003	4,011	8,358	12,118	4	12	24	17	
Wind	342	1,597	1,870	4,604	8,107	10,591	7	13	21	11	
Hydro	3,449	4,343	4,327	5,078	5,890	6,809	15	15	14	1.8	
Bioenergy	341	695	746	1,145	1,540	1,951	3	3	4	4.9	
of which BECCS				4	5	5		0	0	n/a	n/a
CSP	2	14	15	45	166	329	0	0	1	13	11
Geothermal	68	95	97	183	335	458	0	1	1	7.2	5.5
Marine	1	1	1	8	47	96	0	0	0	24	17
Nuclear	2,756	2,673	2,776	3,351	3,897	4,260	10	10	9	2.1	1.5
Hydrogen and ammonia				9	32	44		0	0	n/a	n/a
BECCS			1	5	112	133	0	0	0	21	19
Coal with CCUS			1	5	51	61	0	0	0	21	16
Natural gas with CCUs					61	72	52	47	26	n/a	n/a
Unabated fossil fuels	14,494	14,435	17,436	16,324	14,074	12,562	62	47	26	0.7	1.0
Coal	8,670	9,439	10,291	9,044	7,211	5,822	36	26	12	1.3	1.9
Natural Gas	4,855	6,333	8,552	6,848	6,501	6,558	23	20	13	0.5	0.1
Oil	969	664	682	432	362	312	2	1	1	5.0	2.7

(Continued)

TABLE 6.4 (*Continued*) World electricity sector

	STATED POLICIES SCENARIO (GW)						*SHARES (%)*			*CAGR (%) 2021 TO:*	
	2010	*2020*	*2021*	*2030*	*2040*	*2050*	*2021*	*2030*	*2050*	*2030*	*2050*
Total capacity	5,196	7,849	8,185	11,964	16,458	19,792	100	100	100	4.3	3.1
Renewables	1,343	2,969	3,278	6,707	10,666	13,653	40	56	69	8.3	5.0
Solar PV	39	741	892	3,020	5,573	7,464	11	25	36	15	7.6
Wind	181	737	832	1,830	2,853	3,564	10	15	15	9.2	5.1
Hydro	1,027	1,329	1,358	1,563	1,795	2,027	17	13	10	1.5	1.4
Bioenergy	83	100	173	246	327	406	2	2	2	4	3
of which BECCS				1	1	1		0	0	n/a	n/a
CSP	1	6	7	17	49	90	0	0	0	10	9.2
Geothermal	11	15	16	28	50	66	0	0	0	7	5.1
Marine	0	1	1	4	19	37	0	0	0	18	14
Nuclear	403	415	413	471	545	590	5	4	3	1.5	1.2
Hydrogen and ammonia				3	13	13		0	0	n/a	n/a
Fossil fuels with CCUS		0	0	1	26	33	0	0	0	28	21
Coal with CCUS		0	0	1	10	13	0	0	0	28	18
Natural gas with CCUs					16	20			0	n/a	n/a
Unabated fossil fuels	3,448	4,421	4,462	4,495	4,441	4,196	55	38	21	0.1	0.2
Coal	1,621	2,161	2,184	2,129	1,906	1,583	27	18	8	0.3	1.1
Natural gas	1,389	1,850	1,850	2,074	2,268	2,422	23	17	12	1.3	0.9
Oil	438	427	427	292	292	237	5	2	1	4.1	2.7
Battery storage	1	18	27	270	768	1,296	0	2	7	29	14

the contribution of hydrogen rises from virtually nil in 2021 to only 44 TWh in 2050. Given the current flooding of investment towards hydrogen and its potential for coupling to renewables for energy storage, this too may be an underestimation.

Whether these projections remain valid, or whether nuclear takes a larger share, or hydrogen grows faster than predicted, remains to be seen. In any event if the nuclear industry does double by 2050, it will be a significant contribution to the low-carbon transition.

It is unlikely that fusion will make any contribution in that period, but SMRs might become a feature.

6.3.12 Current-Generation Nuclear Power Plants

Nuclear (fission) is an established industry with China pushing ahead with 21 new plants, and at least three large stations are scheduled to be built in the UK (Hinkley Point, Sizewell and Bradwell). Elsewhere, new build projects are underway in France, Finland, UAE, Egypt, Belarus, Iran, Russia, India, Slovakia, Korea, Bangladesh, Argentina, Brazil, Turkey and the USA, scheduled by 2030. Several countries have also uprated their existing plants.

Thus, the forecast growth in nuclear is happening. There is an established suite of available reactor designs to choose from. Global supply chains are established or being established, for both components and expertise. There are established design codes which are used internationally, and national regulatory bodies have been established or strengthened. Thus, the current situation appears stable and able to sustain the current growth.

However, disposal of HLW remains a crucial issue and could hamper future expansion if not addressed within this generation. Apart from disposal sites already identified, it seem that construction work cannot be started at new sites by 2050, but it may be possible for other sites to be identified and proving trials initiated within this period. Expanded use of dry storage for SNF may be an interim solution until sites become available later this century or even beyond.

Into the future, it is likely that reactor designs will continue to be simplified, standardised and made less expensive to build.

6.3.13 Integration with Hydrogen

With the development of proton exchange membrane (PEM) technology for electrolysers and hydrogen fuel cells, there has been a huge growth in hydrogen production and new technologies are being developed for the use of hydrogen in transportation, heavy industry and domestic usage.

HYBRIT (Hydrogen Breakthrough Ironmaking Technology) – a collaboration between steel producer SSAB, mining company LKAB and Vattenfall – started in 2017 and is aiming to transform the steel making process. Thus, the hydrogen industry is expanding rapidly. The driver for much of this is the effective use of intermittent renewable energy with hydrogen acting as a storage medium available to meet demand when renewables are not available and as a gas alternative to electricity.

Classic electrolysis is the process of splitting of water into hydrogen and oxygen by passing electricity between an anode and cathode submerged in an electrolyte. In the PEM process, the electrolyte is replaced by the PEM which separates the anode from the cathode and allows hydrogen protons to penetrate to the cathode where hydrogen is collected, and oxygen at the anode. This process is commercially attractive using electricity from renewable sources such as wind and solar. Large parts of the world with high solar energy may be able to produce hydrogen as an export commodity. Electricity for electrolysis may come from anywhere, including nuclear. UKAEA gives an extensive discourse on the technologies for hydrogen production using nuclear energy (UKAEA, 2024).

A paper by Grant in *Nuclear Future* (2013), postulated a small town with a surrounding hydrogen ring main, with the power for electrolysis coming from a SMR, thus providing electricity, gas and energy storage capacity. One can see variants of this being possible especially in areas with little renewable energy resources or urban areas where space for renewables is not possible.

6.3.14 End of Fission

Currently nuclear power from fission is expanding and is likely to continue to do so beyond 2050, provided there is not another major nuclear accident like Fukushima or political unrest that threatens the security of a nuclear power plant. Such occurrences would likely undermine public opinion about nuclear. It appears also that waste disposal concerns will not be solved in the short term, and so security and environmental concerns are likely to grow.

SMRs (including liquid fluoride thorium reactors) are likely to become more common, as they should be easier to license and less capital-intensive to build and may be linked to future growth in a hydrogen market.

There may also be a tipping point where the cost of mining new uranium or use of reprocessed materials becomes more expensive compared to next-generation renewables and emerging fusion power, and this could undermine the economic viability of fission or support the use of thorium fuel.

These are the issues that will affect the sustainability of nuclear power as it is today, and it may not be viable into the next century, although thorium fuel may prove viable.

6.3.15 Expansion of Fusion

At the same time, hopes are high for the emergence of fusion as a future sustainable energy source. However, the technology level required for fusion is considerable and will require ongoing investment in R&D over the next few decades, and in skills and training, but given the experience of a vast increase in technology generally over the past century, and its continuing growth, it seems likely that a viable fusion technology/industry will emerge.

6.3.16 Nuclear Power: Conclusions

As described above, the current nuclear fission power industry is very well established all over the world and involves a sophisticated supply chain covering the whole fuel cycle from uranium mining to waste disposal and including design and manufacture of complex equipment and components. This supply chain relies on raw materials and expertise continuing to be available at an affordable price, which itself relies on financial and political stability.

The sustainability of nuclear power relies on these current conditions continuing, at least until the waste disposal issue is dealt with satisfactorily. If the nuclear industry were to stop right now, not only would we lose 10% of the world's electricity but we would also be left with a backlog of nuclear waste especially SNF and HLW, with all the associated safety concerns. So, it is imperative that nuclear does remain sustainable at least until the end of this century to allow these concerns to be addressed.

In their extensive report on the nuclear fuel cycle, Massachusetts Institute of Technology has the following conclusions aimed at policymakers in the USA but relevant elsewhere (MIT, 2011):

1. For the next several decades, light water reactors using the once-through fuel cycle are the preferred option for the USA.
2. Planning for long-term interim storage of SNF – for about a century – should be an integral part of fuel cycle design.
3. For the longer term, there are multiple viable fuel cycle options with different economic, waste management, environmental, resource utilisation, safety and security, and non-proliferation benefits and challenges. A significant research agenda is needed to explore, develop and demonstrate the advanced technologies to the point of allowing informed future marketplace and policy choices.

Thus, the question is not whether nuclear is sustainable or not, but how best it can be made to contribute to development of a sustainable energy future for the world.

Fission has many years to run yet and is already a low-carbon alternative to fossil fuels. Improvements to the current generation of large nuclear power plants are continuing to enhance their efficiency and safety, and new SMRs are gaining acceptance with the potential for linkages to hydrogen production as part of continuing diversity of energy supply and supporting the development of renewables.

Practical application of fusion appears to still be some years away but with development of DEMO to follow-on from ITER and new prospects such as development of spherical tokamak technology such as STEP or MTF concepts, it is quite possible that fusion power plants will be operational this century. With fusion, the waste disposal issue does not exist to anything like the same extent of fission, and provided the outstanding technical issues can be resolved, fusion offers a sustainable option into the far future. But all this requires stable financial conditions and availability of specialist raw materials and resources to continue.

Beyond the Brundtland definition of sustainability, it is contended that four conditions are needed:

1. Developments must make things better today – bringing energy to more people, combating climate change, addressing energy security concerns, etc. If it does not make things better for the current generation, then it may not be possible to get support for the development, despite potential future benefits.
2. Developments must be efficient in the use of resources. If they are not sufficiently efficient, then other options will be preferred. For instance, if a particular form of fusion proves overly complex or too dependent on scarce resources, then it may be overtaken by another method or indeed by renewable developments.
3. There must be banking for the future. This is the corollary of item 1. People are all too aware of the problems of short-term planning and see the current climate crisis as having been caused by not thinking about future consequences. If scarce resources are to be employed in nuclear, then the investment in time, money and talents must give back to future generations – i.e. start to create the sustainable energy future, including taking away the current nuclear waste problem.
4. Finally, there must be visible fairness. Society will no longer tolerate unfairness, where the rich get richer and the poor get poorer, where the developed world prospers at the expense of the developing world or where a sustainable future is for some and not others. Along with ideological differences, visible unfairness is a great danger to world instability.

Many dystopian views of the future question whether technological development will indeed be able to continue as described above, or will the whole system regress in some way. Certainly, there are many current dangers of political and social unrest that could cause a hiatus in the road to a sustainable energy future.

In conclusion, provided political, economic and social stability can continue, it is contended that nuclear can contribute to a sustainable energy future.

6.4 RENEWABLES

6.4.1 General Introduction to Renewable Energy

In the 1970s, although global warming had not yet been fully recognised yet as a risk, there was an increasing spotlight on pollution, and fossil fuels, particularly coal, as the likely culprit for the problem This drew attention to finding clean or renewable forms of energy. Renewable energy is derived from natural sources that are replenished at a higher rate than they are consumed. These sustainable sources include principally hydro, sunlight and wind, with additionally marine (tide and waves), geothermal and biofuels also becoming significant (EDF, 2024).

The status of where the world was in 2021 on a path to achieve net zero by 2050 or even 2070 (the later date as requested by China and India), was 38% renewables and nuclear, 37% coal, and the balance being gas at 25%. The general balance of renewable energies is illustrated in Figure 6.2 (Ritchie et al., 2024). Although China is broadening its scope of type of development within the energy sector, it has invested a significant amount in coal-fired power stations, particularly since about 2005, and produces more than 50% of the world's coal-based output as a result. It does not seem likely that the net zero target will be met by mid-century, the figure strongly recommended by the UN, although the efforts that are being made are showing a gradual change in the overall picture.

Hydropower is currently the largest renewable source of electricity globally – though that is expected to change by the end of this decade. Growth in capacity is forecast to remain stable through 2030, driven by additions in China, India, Africa and Southeast Asia. Although hydropower generation is expected to increase globally as new projects become operational, the technology's share in total power generation is expected to decline as growth in electricity demand outpaces deployment. China is expected to account for 60% of the renewable's expansion in global capacity to 2030. China's cumulative solar capacity has almost quadrupled, and wind capacity has doubled, driven by cost-competitiveness and supportive policies.

6.4.2 Hydropower Description

Hydropower, or hydroelectric power, is a renewable source of energy that generates power derived from the difference in head of flowing water from a higher to a lower level. This can be from rivers or man-made installations, where water flows from a high level. There are many types of hydropower facilities, though they are all powered by the kinetic energy of flowing water as it moves downstream. Hydropower utilises turbines

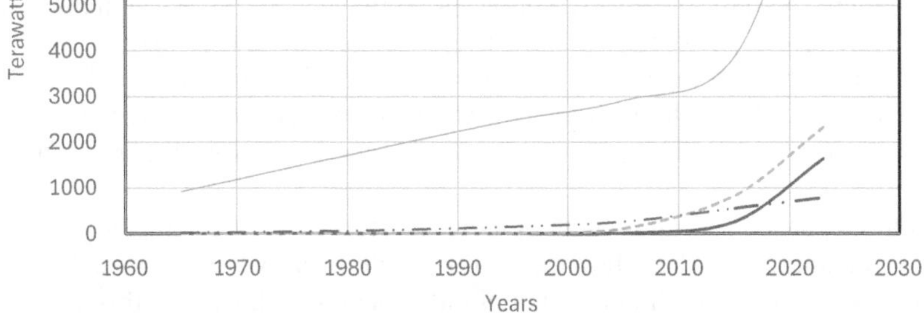

FIGURE 6.2 Changing picture of renewable energy.

and generators to convert that kinetic energy into electricity, which is then fed into the electrical grid to power homes, businesses and industries.

IFC (2021) describes hydropower as a crucial power supply option for several reasons:

- It is a renewable energy resource that can contribute to sustainable development by generating local, typically inexpensive power.
- Hydropower reduces developing country reliance on imported costly fuels.
- Hydro systems can offer additional benefits including water storage for drinking and irrigation, drought-preparedness, flood control protection, aquaculture and recreational opportunities.

Hydropower is one of the oldest and largest sources of renewable energy, with roots tracing back thousands of years, when it was initially used for powering flour mills or distributing water for irrigation schemes. Modern hydropower schemes take many forms with three main configurations (GOV.uk, 2013):

Run-of-river schemes: These schemes use the natural flow of a river, although the addition of a weir can enhance the continuity of the flow. These schemes can use a diversion, where water is channelled from the river to a remote powerhouse containing the turbine and generator. The problem with most run-of-river systems without upstream storage is the large variation in flows experienced in a river between wet periods and droughts. The issue becomes even more critical when climate change rainfall unpredictability is taken into account.

Dam storage schemes: In storage schemes, a dam impounds water in a reservoir that feeds the turbine and generator sometimes located within or close to the dam itself. Alternatively, these schemes commonly use a headrace tunnel for diverting the water from the reservoir to the powerhouse at a lower level. Figure 6.3 shows the layout of the Trang Son hydropower site in the mountains of northwest Vietnam. This typical medium-sized project costs around US$ 410 million and can deliver 260 MW to the national grid (Figure 6.3).

Pumped storage: These schemes incorporate two reservoirs at different elevations that can generate power as water moves down from one to the other (discharge), passing through a turbine. The turbines are reversed and act as pumps to fill the upper reservoir when energy is in excess of requirement and has a low tariff. When there is a peak demand, water is released to generate energy through the turbines to generate electricity when the tariff is higher (IFPSH, 2021). Although not considered a renewable energy (because of its reliance on electricity), pumped storage is considered overall energy-efficient.

Essentially hydropower requires an appropriate river and a topographic environment (hilly or mountainous) that allows a significant drop in hydraulic head. Norway, for example, which has a small population of only about 5 million, uses this as a primary source for their electricity requirement. China, however, is by far and away the largest producer at 14.4% of the world's total. In addition, North America, USA and Canada,

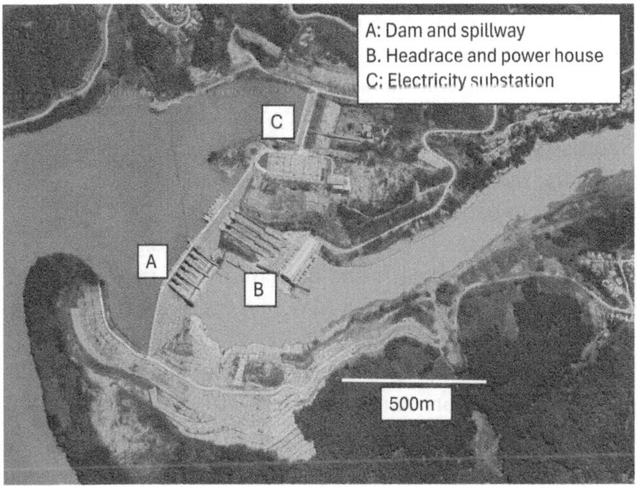

FIGURE 6.3 Trung Son power station, Vietnam (Google Earth image).

together rank second to China, although significantly less. Scotland, a small user of hydropower, made significant investment in hydro over a 25-year period from the mid-1930s. Many other countries around the world also have built hydropower projects. Indonesia, provides good examples of a variety of hydro projects. Smaller Asian countries such as Nepal and an Laos with ideal physical environment produce power for export to larger neighbouring countries.

6.4.3 Hydropower: Some Examples

Some of the largest run-of-river schemes are in Indonesia where rivers are supplied from upstream by high-level volcanic lakes. A major system is a cascade of three schemes on the Asahan River, in Sumatra, supplied by Lake Toba giving reliable flows throughout the year of between 120 and 126 m^3/s. The Sigura Gura has a head of 231 m and electricity capacity of 286 MW, the downstream Tangga station has a head of 250 m and capacity of 317 MW, whilst a third under construction will have a 175 MW capacity. Their intakes have diurnal storage that allows supply to mirror demands over a 24-hour day. There is also the Jatiluhur scheme which involves three dammed steps of the Citarum River, east of Jakarta. This not only has an important prime role to supply Jakarta with water but also has a secondary role of providing electricity.

The UK has some 440 large dams (higher than 15 m) that could be fitted with hydropower units. However, multi-purpose schemes often have conflicting operations with the hydro operators trying to maintain as high a level as possible in the reservoir to maximise power production, while other users want to run down the active storage over the driest parts of the year or leave substantial space in the reservoir to absorb floods.

In the UK, there has traditionally been excess energy at night as the thermal stations on the grid are not closed down and a 6-hour peak period between 3 and 11 pm where systems that can respond quickly to rapid increases in demand are needed. The first UK pumped storage hydro (PSH) the Cruachan or Loch Awe scheme, was brought on stream in 1963. The Welsh located Dinorwig pumped storage, which was being designed in parallel with Cruachan, started operating in 1964, is rated at 1.4 GW with four 350 MW turbines and is considered the most cost-effective power plant in the UK. Scotland has the most suitable sites for development, and this has been recently recognised by the Scottish Government, with new initiatives summarised below.

PSH is by far the most predominant large-scale energy storage technology worldwide. More than 97% of worldwide storage capacity was PSH by 2015, more than 23 GW of additional capacity are currently under construction and another 8.3 GW have been announced to be constructed in the near future. In the UK, there are four PSH facilities planned which amount to 2,828 MW of total capacity, as follows:

1. The announcement made by Scottish Power in early 2016 of their intention to upgrade the Cruachan pumped hydropower station, part of the Loch Awe project in order to double its power output; the first operating phase was completed in 1964.
2. The proposed new pump storage schemes, Coire Glas (1,500 MW), which was approved by the Scottish Government in October 2020, Balmacaan (600 MW) and Glyn Rhonwy (100 MW).
3. The planned Sloy power station conversion into a pumped storage facility (adding 60–252.5 MW).
4. Two new PSH schemes recently announced that are under planning: Muaitheabhal2 (300 MW) and Glenmuckloch3 (105 MW).

Other existing schemes in the UK are Foyers in Scotland (300 MW), and Ffestiniog in North Wales (360 MW) built in 1963. Other intentions by the ILI Group include the initial planning phase for the 1.5 GW Balliemeanoch PSH project at Loch Awe, the 450 MW Red John project at Loch Ness (awarded planning consent from Scottish Ministers in June 2021) and submitting a planning application for the 600 MW Corrievarkie project at Loch Ericht in 2022.

Many UK reservoirs are only used during the dry season for water supply when energy demands are low but have little use during the winter and spring months when most are full. Other water bodies have little human usage. Climate change predictions in the hillier parts of Britain are for 40% winter increases in runoff by 2050, when electricity demands are highest. Where there are two existing and nearby storages at different levels, they need to be connected by a pressure conduit, flow control systems and a power station that can be hidden underground with little visual impact. In addition, such projects offer a lot of interesting experience for British civil, electrical and mechanical engineers and can supply much needed skilled employment.

6.4.4 Solar

Solar radiation is essentially a free resource available to a greater or lesser extent worldwide. Solar photovoltaics (PV) power plants convert solar radiation into electricity. Over the last decade, the solar power sector has seen installation costs fall dramatically and global installed capacity rise massively.

Around 20% of the global population lives in 70 countries boasting excellent conditions for PV, where long-term daily output averages exceed 4.5 kWh/kWp. Countries in the Middle East and North Africa region and Sub-Saharan Africa dominate this category, accompanied by Afghanistan, Argentina, Australia, Chile, Iran, Mexico, Mongolia, Pakistan, Peru and many nations of the Pacific and Atlantic islands. At the lower end of the ranking, 30 countries accounting for 9% of the global population score an average output below 3.5 kWh/kWp, dominated by European countries and also including Ecuador and Japan. Finally, countries in the favourable middle range between 3.5 and 4.5 kWh/kWp account for 71% of the global population. These include five of the six most populous countries (China, India, the USA, Indonesia and Brazil) and 100 others (Canada, the rest of Latin America, southern Europe and African countries around the Gulf of Guinea, as well as central and southeast Asia).

In the current era of global climate change, PV technology becomes an opportunity for countries and communities to transform or develop their energy infrastructure and step up their low-carbon energy transition (World Bank, 2020). The global pattern of use (theoretical PV potential) is determined mainly by latitude, occurrence of clouds, air temperature terrain elevation and shading, atmospheric aerosol concentration and atmospheric moisture content (IEA, 2020). For example, Indonesia has been assessed as being able to produce 80 GW of solar power, which could theoretically provide for much of its current energy needs, although unlikely (Younger et al, 2022).

Solar PV equipment can be manufactured in large plants, which creates economies of scale, but can also be deployed in very small quantities at a time. This allows for a wide range of applications, from small residential roof-top systems up to utility-scale power generation installations. Solar PV can be the least costly option for new electricity generation in a significant majority of countries (IEA, 2024). There are two main applications for solar PV, on the roofs of properties (Figure 6.4) and in multifaceted ground arrays covering a large area. Solar has additional use for the batteries of electric

FIGURE 6.4 Solar panel roof array.

FIGURE 6.5 Household battery charging in rural environment, Bangladesh.

motor vehicles (EVs), and there is a slow but steadily increasing penetration of EVs in this very large potential market for motor vehicles. Figure 6.5 shows an example of the small-scale use of solar energy in a rural environment in Bangladesh.

6.4.5 Wind

The renewable energy agenda in UK has really picked up in the past two plus decades. UK is particularly suitable for wind energy being an offshore island and gets much wind mostly and steadily from the Atlantic Ocean where the trade wind blows from west to east, dictated largely by the spin of the earth around the sun. It is also successfully harnessed in the countries with an east Atlantic coastline, namely Ireland, Germany, The Netherlands, Belgium, France, Portugal and Spain, as well as would be by countries from North Africa facing the Atlantic.

Wind energy is divided into that which is garnered by wind turbines on land and that which is taken offshore. This latter may be further divided into that using solid foundations for the turbines and that using large floating turbines. In the early days, a wind turbine capacity was modest, but nowadays the turbines can be large, e.g. 10 MW capacity.

While Britain can continue to add to its wind resources, both on land and offshore, there has been growing reluctance for further wind farms at land-based sites. Whatever there is plenty of additional capacity. The drawback for wind energy is there is no power produced when there is no wind. Wind energy has to be integrated into a power system that is flexible enough to alternate between different energy sources, whether clean or fossil fuel-based.

Recently, marine environmentalists have drawn attention to the foundations of offshore wind turbines disturbing whales, in particular. The whales' in-built sonar system no longer works where they have to pass through these foundation forests, and they find themselves misdirected. In addition, environmentalists also tend to complain as birds, especially seabirds, can be cut down by the blades that drive the wind turbines.

Wind turbines are expensive to design and install. This is particularly the case of offshore turbines where extensive and expensive investigations are necessary for either the turbines themselves or the many anchors required for the floating option.

Once built, wind farms are economical and cheaper to run than fossil fuel alternatives. Their return on investment makes them viable for locations where there is reliable wind capacity, particularly since the price of oil was quadrupled by OPEC in 1973. It is likely that Britain will increase its investment in wind power in the coming period, as it has become more confident, with usage of the operational needs. At the moment, the UK's provision of energy is 80% by renewables, which is in sharp contrast to that of two decades earlier.

6.4.6 Marine Energy

Tidal energy is produced by ocean water surges during the rise and fall of tides. Tidal energy is also a renewable source of energy. During the 20th century, engineers developed ways to use tidal movement to generate electricity where there is a significant tidal range – the difference in area between high tide and low tide. All methods use special generators to convert tidal energy into electricity. Tidal energy production is still in its infancy. The amount of power produced so far has been small. There are very few commercial-size tidal power plants operating in the world.

For most tidal energy generators, turbines are placed in tidal streams of fast-flowing body of water created by tides. flowing through them. Because water is much denser than air, tidal energy is more powerful than wind energy. Unlike wind, tides are predictable and stable. Where tidal generators are used, they are placed in mid-stream. However, placing them in tidal streams is complex, because the machines are large and disrupt the tidal flow they are trying to harness. The environmental impact could be severe, depending on the size of the turbine and the site of the tidal stream. Turbines are most effective in shallow water, which produces more energy and allows ships to navigate around the turbines. A tidal generator's turbine blades also turn slowly, which helps marine life and avoids it getting caught in the system.

The world's first tidal power station was constructed in 2007 at Strangford Lough in Northern Ireland. The turbines are placed in a narrow strait between the

Strangford Lough inlet and the Irish Sea. The tide can move at 4 m/s across the strait. Another scheme is currently well under development, MeyGen, in the north of Scotland, located between the mainland and the offshore small island of Stroma, across the Pentland Firth. This has the ultimate aim of generating some 50 GW of hydropower.

Another type of tidal energy generator uses a large dam or barrage. Turbines inside the barrage harness the power of tides the same way a river dam harnesses the power of a river. The barrage gates are open as the tide rises. At high tide, the barrage gates close, creating a pool, or tidal lagoon. The water is then released through the barrage's turbines, creating energy at a rate that can be controlled by engineers.

The environmental impact of a barrage system can be quite significant. The land in the tidal range is completely disrupted. The change in water level in the tidal lagoon might harm plant and animal life. The salinity inside the tidal lagoon lowers, which leads to changes in the organisms that are able to live there. As with dams across rivers, fish are blocked into or out of the tidal lagoon. Turbines move quickly in barrages and marine animals can be caught in the blades. With their food source limited, birds might find different places to migrate.

A barrage is a much more expensive tidal energy generator than a single turbine. Although there are no fuel costs, barrages involve more construction and more machines. Unlike single turbines, barrages also require constant supervision to adjust power output.

The tidal power plant at the Rance River estuary in Brittany, France, uses a barrage. It was built in 1966 and is still functioning. The plant uses two sources of energy, tidal from the English Channel and river current energy from the Rance River. The barrage has led to an increased level of silt in the habitat. Native aquatic plants suffocate in silt, and a flatfish called plaice is now extinct in the area.

The largest facility worldwide is the Sihwa Lake Power Station in South Korea, which also has the next largest. The USA has no tidal plants and only a few sites where tidal energy could be produced at a reasonable price. China, France, Britain, Canada and Russia have much more potential to use this type of energy. Engineers are working to improve the technology of tidal energy generators to increase the amount of energy they produce.

A tidal energy generator using tidal lagoons would function much like a barrage. Unlike barrages, however, tidal lagoons can be constructed along the natural coastline. A tidal lagoon power plant could also generate continuous power. The turbines work as the lagoon is filling and emptying.

The environmental impact of tidal lagoons is minimal as the lagoons. The lagoons can be constructed with natural materials, for instance rock. They would appear as a low breakwater (sea wall) at low tide and be submerged at high tide. Animals could swim around the structure, and smaller organisms could swim inside it. Large predators, like sharks, would not be able to penetrate the lagoon, so smaller fish would probably thrive and seabirds, that feed on fish, would likely flock to the area.

6.4.7 Biofuels

Biofuels include solid biomass (excluding for traditional uses), liquid biofuels, biogases, electrolytic hydrogen made from renewable electricity (renewable hydrogen) and e-fuels (fuels made from renewable hydrogen, including e-kerosene, ammonia and methanol). Renewable fuels are garnering increasing interest as an option to reduce GHG emissions in sectors that are difficult to electrify, while also providing energy security and economic development opportunities.

Biofuels play a particularly important role in decarbonising transport by providing a low-carbon solution for hard-to-abate sectors such as trucking, shipping and aviation. They can often be used in existing engines with little to no modification (IEA, 2024).

Most biofuel production currently uses so-called conventional feedstocks, such as sugar cane, corn and soybeans. Expanding biofuel production to advanced feedstocks is critical to ensuring minimal impact on land use, food and feed prices and other environmental factors.

Countries and regions making notable progress to boost biofuels include the following:

- India achieved 10% ethanol blending in 2022, ahead of schedule, in its pursuit of a 20% blending target by 2025.
- Brazil is planning to increase biodiesel blending to 15% by 2026, up from 10% in 2022.
- The United States Inflation Reduction Act (IRA) provides production and investment support for biofuels estimated at USD 9.4 billion to 2031.
- Canada is implementing its Clean Fuel Regulations in 2023, which require a 13% reduction in GHG emissions intensity for transport fuels by 2030.
- The European Union is approaching agreement on its updated Renewable Energy Directive (RED III), which would double the requirements for renewables content in transportation fuels, including biofuels, compared to existing targets.

6.4.8 Geothermal

Geothermal energy is stored in the form of heat beneath the surface of the earth. It is a reliable and constant source of low-carbon, renewable heat that is not dependent on weather conditions. It is available at depths from a few metres to several kilometres from where it can be extracted using different technologies. The amounts and the accessibility of this energy are to a large extent governed by the geology and structure of the earth (Chapter 4). A number of countries have good reliable resources of geothermal energy, particularly those that have boundaries close to tectonic plates or other major structural faults.

The country which generates the most power from geothermal sources is the USA, based on outlets depending on the N. American plate, assessed at nearly 4 GW. The next is Indonesia, 2.4 GW, followed by Philippines, Turkey, New Zealand and Kenya. While it is worthwhile obtaining the renewable power from geothermal sources, it is small in terms of the total energy that is generated, albeit in terms of Iceland which takes all its small amount required from this one source.

Two countries that make much use of geothermal are Iceland and Indonesia, especially the former which takes nearly all its energy needs from geothermal. Both lie close to major plate boundaries of different type; Iceland is split by a constructive boundary and Indonesia (Sumatra and Java) adjacent to a destructive plate boundary.

In the UK, the temperature of geothermal heat and extraction technology used varies by depth and geology. Near the surface, shallow geothermal heat is stored in the ground, aquifers and in flooded coal mine workings. It can be extracted by ground source heat pumps and is suitable to be used for heating. Deep geothermal energy is extracted from depths greater than 500 m, and it can be suitable for either direct-space heating or power generation depending on temperature and depth.

Sweden and the Netherlands have ambitious programmes for expanding geothermal use. France and Germany are also looking to expand their geothermal energy usage and have developed ambitious targets and strategies to increase its use. As France and Germany have similar geology to the UK, they are presented as potential examples to follow.

6.5 ROUND-UP OF RENEWABLE AND CLEAN ENERGY SOLUTIONS

While hydro is expected to be eventually overtaken by wind and solar, it will continue to play a key role as a dispatchable power source to back up variable renewables, although increasingly erratic rainfall due to climate change is also disrupting hydro production in many parts of the world. Pumped storage could also potentially play a major role in balancing out variations in solar and wind generation.

China continues to lead in terms of capacity additions, with 8 GW added in 2023, equal to two-thirds of all global growth. Hydropower remains an important part of the 14th Five-Year Plan for Renewable Energy released in 2022, but capacity additions in 2023 and in the coming years are expected to remain significantly below levels observed in the past decade due to a diminishing number of suitable sites and environmental constraints.

India is continuing to develop several large hydropower projects, with significant capacity expected to come online in the coming years. Hydropower is one of the crucial technologies for fulfilling a commitment to reach 500 GW of non-fossil electricity capacity in 2030.

It is considered that while renewables or nuclear and even hydro will receive much attention in terms of development and provide most of our energy requirements going forward, there will be a smaller demand over the remainder of this century. A guideline target for maximum use of fossil fuels would seem to be below a level that was used pre-1960s, which indicates that emission levels should be not much more than pre-industrial ones, when no obvious changes were taking place to the climate.

Some countries have taken serious measures towards halting increasing carbon emissions if not reversing the unwanted trend, and the signs are there that they have been successful. Other countries have still to join, taking note that a number of smaller, relatively undeveloped countries hardly commit any carbon emissions to the total and can carry on as usual, but noting that they can also make use of the knowledge that has been gained by their predecessors.

Developments in different ways to harness solar power will be taking place, particularly through the US Space Program as well as research on, for instance, hydrogen. Overall, the lesson of today, will be to find more efficiency whatever mode is used, assuming not fossil fuels, and to minimise any waste and pollution that is caused.

Finally, attention should be paid to the Koppen–Geiger zones of world climate (see Chapter 3), to decide on which energy modes are most appropriate for each area, noting again the continuing steady stream of developments in Asia and notable expansion in population and with consequent need for energy in Sub-Saharan Africa. These countries have the opportunity to learn from more developed countries when trying to bring about changes towards renewable forms of energy generation pertinent to their needs and location in the world.

Looking into the future, the sun can be expected to provide at least half of mankind's energy requirements, probably more. The US Space Program, as one of its briefs, is carrying out research in how to harness the sun's radiated energy for domestic use as well as for defence requirements. It is quite possible that the sun's energy, which is a permanent sustainable presence, will be harnessed more directly but it is not a short-term answer. The renewable and clean energy solutions that have been developed will probably be relevant until the end of the century and into the next for all humankind's needs, although it is expected that fossil fuels will still be required for some uses.

NOTE

1 A tokamak uses powerful external magnetic fields to confine and control the hot plasma of fusion fuels in a ring-shaped container called a 'torus'.

REFERENCES

EDF (2024) *What is Renewable Energy*. Available at: https://www.edfenergy.com/energywise/renewable-energy-sources

EDF (2025a) *Hinkley Point B*. Available at: https://www.edfenergy.com/energy/power-stations/hinkley-point-b

EDF (2025b) *Hinkley Point C*. Available at: https://www.edfenergy.com/energy/nuclear-new-build-projects/hinkley-point-c

General Fusion (2025) *Magnetised Target Fusion*. Available at: https://generalfusion.com/fusion-technology/

GOV.uk (2013) *Harnessing Hydroelectric Power*. Available at: https://www.gov.uk/guidance/harnessing-hydroelectric-power

Grant, P. M. (2013) Energy for the city of the future. *Nuclear Future* 1(1): 17–21.

IAEA (2021) *China HLW Repository*. Available at: https://www.iaea.org/newscenter/news/china-begins-construction-of-its-first-underground-research-laboratory-for-high-level-waste-disposal

IAEA (2024) *Nuclear Power Reactors in the World*. International Atomic Energy Agency, Vienna. Available at: https://www-pub.iaea.org/MTCD/Publications/PDF/p15748-RDS-2–44_web.pdf

IFC (2021) *Hydroelectric Power. A Guide for Developers and Investors*. World Bank Group. Fichtner Management Consulting, Stuttgart, Germany. Available at: https://ppp.worldbank.org/sites/default/files/2021-10/Hydroelectric%20Power%20-%20A%20Guide%20for%20Developers%20and%20Investors.pdf

IFPSH (2021) *International Forum on Pumped Storage Hydropower*. Executive Summary, p. 9, Available at: https://cdn.prod.website-files.com/64f9d0036cb97160cc26feba/64f9d0036cb97160cc271277_IFPSH%20-%20Executive%20Summary%20-%2016%20Sept.pdf

International Energy Agency (2021) *World Energy Balances:* Overview. IEA, Paris. Available at: https://www.iea.org/reports/world-energy-balances-overview

International Energy Agency (2022) *World Energy Outlook 2022*. Available at: https://iea.blob.core.windows.net/assets/830fe099-5530-48f2-a7c1-11f35d510983/WorldEnergyOutlook2022.pdf

IEA, (2024) *World Energy Outlook*. International Energy Agency. Available at: https://iea.blob.core.windows.net/assets/140a0470-5b90-4922-a0e9-838b3ac6918c/WorldEnergyOutlook2024.pdf

IPCC (2010) *Climate Change: The IPCC 1990 and 1992 Assessments*. Available at: https://www.ipcc.ch/report/climate-change-the-ipcc-1990-and-1992-assessments/

ITER (2025) *What is ITER*. Available at: https://www.iter.org/proj/inafewlines

LLNL (2025) *What is National Ignition Facility*. https://lasers.llnl.gov/about/what-is-nif

Martin, R. (2012) *Superfuel: Thorium, the Green Energy Source for the Future*. Palsgrave-Macmillan, Basingstoke.

MIT (2011) *The Future of the Nuclear Fuel Cycle*. Available at: https://energy.mit.edu/publication/future-nuclear-fuel-cycle/

NLF (2025) *What Is the National Ignition Facility*? Lawrence Livermore National Laboratory, California, USA. Available at: https://lasers.llnl.gov/about/what-is-nif

Paxman, J. (2022) *Black Gold*. HarperCollins Publishers, London.

Power Technology (2016) Espejo de Tarapaca Project. https://www.power-technology.com/projects/espejo-de-tarapaca-project/

Ritchie, H. (2019) *The death of UK coal in five charts*. Published online at OurWorldinData.org. Available at: https://archive.ourworldindata.org/20251125-173858/death-uk-coal.html

Ritchie, H. and Roser, M. (2017) *A Sense of Units and Scale for Electrical Energy Production and Consumption*. Our World of Data. Electricity Usage. Available at: https://ourworldindata.org/scale-for-electricity

Ritchie, H., Roser M. and Rosado P. (2024) *Renewable Energy*. OurWorldinData. Available at: https://ourworldindata.org/renewable-energy

UKAEA (2024) *STEP*. Available at: https://storymaps.arcgis.com/stories/2f7cd4b0ce984a5595e7d26e8a1dca47

USGS (2024) *What are the Types of Coal*. United States Geological Survey. Available at: https://www.usgs.gov/faqs/what-are-types-coal

Wikipedia (2024) *Fusion Power*. Available at: https://en.wikipedia.org/wiki/Fusion_power#:~:text=Fusion%20power%20is%20a%20proposed,are%20known%20as%20fusion%20reactors

WNA (2025) *Nuclear Fuel Cycle Overview*. World Nuclear Association. Available at: https://world-nuclear.org/information-library/nuclear-fuel-cycle/introduction/nuclear-fuel-cycle-overview

World Bank (2020) *Global Photovoltaic Power Potential*. ESMAP report. Available at: https://documents1.worldbank.org/curated/en/466331592817725242/pdf/Global-Photovoltaic-Power-Potential-by-Country.pdf

Younger, J. S., Sumiarso, L., Murray, M. I., Parry, D. E. and Prisandani, U. Y. (2018) Energy resources of Indonesia and their application. In *2nd Dundee Energy Forum: Evolving Global Energy and Resources Landscape*. University of Dundee, Scotland.

Water

7

Scott Younger and David Meigh

7.1 HISTORIC CONTEXT

Life, in all its forms appeared about 3.7 billion years ago with hominids (man's direct ancestors) evolving about 2.3 million years ago. Java man appeared about 1.6 million years ago with his grave marked in Central Java. Humanity as we understand and know it, is much more recent, a mere few thousand years. None of this development would have occurred without the presence of water, which originally had made all life possible.

Water is undoubtedly the most precious resource on the planet, and the UN (2022) forcefully puts the case for its importance as follows:

> Water and sanitation are at the core of sustainable development, and the range of services they provide underpin poverty reduction, economic growth and environmental sustainability. However, in recent decades overexploitation, pollution, and climate change have led to severe water stress in locales across the world.

Effective sustainable human development has from earliest times been associated with the availability of freshwater, for example, in Mesopotamia (Tigris and Euphrates) and Egypt (Nile), where not immediately available water was transported by canal or aqueduct, as in developing the Roman empire where aqueducts were a common feature throughout. Closer to home, the first reported example in UK of taking freshwater from source (Loch Katrine) by a combination of tunnel and aqueduct was to supply the city of Glasgow in 1859. The works involved 21 km hard rock tunnelling within a total length of 42 km (Paxton, 2009). This action confirmed important evidence that the dangerous disease of cholera was not a miasma, as commonly believed, but water-borne, and led to a profound change in medical thinking, and underlined the critical importance of a sustainable supply of water to urban communities across the world.

It is clear we could not survive without freshwater, but it is not always treated with the 'respect' it is due. Nor is it evenly distributed across the planet with particular problems occurring, the lack of freshwater is where the demand is most significant. Water is used for multifarious purposes and, consequently, this brief appreciation can only cover the main

DOI: 10.1201/9781003566274-7

points, focussing on water sector challenges to sustainable development. More detailed reviews of the water sector can be found in Feldman (2012) and Lane et al. (2017).

7.2 THE WORLDWIDE DISTRIBUTION AND USE OF WATER

7.2.1 Overview

There is a lot of water on planet Earth, but by far most of it goes to make up saline oceans that have served the sea transport of people and goods, leisure activities, the fishing industry, and for the development of trade. Seawater is not usable directly to satisfy the requirements of human consumption and hygiene. Table 7.1 shows the breakdown of how water is divided up and where it can be found.

Sir David Attenborough has spent most of his life in covering life on the planet and making us more aware, quite spectacularly, of it, as well as our total responsibilities to all life on it. Over the past decade, he has turned his attention to research into our oceans, making us aware that significant pollution damage is being caused to seabeds, particularly by current commercial fishing methods. Changes are required here (Attenborough and Bitfield, 2025).

Sustainability and development of the water sector have now become priority issues of the world's international lending agencies following the UN meeting held in New York at UN headquarters in 2012 (UN, 2012) where all parties realised that availability of and ready access to water is complex and raises many issues. Much earlier a 1983 report entitled *Our Common Future* published by the UN's "World Commission on Environment and Development" said over four decades ago, that "major, unintended changes are occurring in the atmosphere, in soils, in waters, among plants and animals, and in the relationships among all of these" (UN, 1987). This was an important wake up call, which it has taken a long time for people to digest. Observation of how water is treated across the world places a doubt that the warnings in that report have been properly heeded.

The population is expected to increase by over 2 million in the decade, 2020–2030, and underground freshwater resources to diminish further. The UN world water

TABLE 7.1 Worldwide water

S. NO.	*SOURCE*	*PERCENTAGE (%)*
1	Oceans (saltwater)	97.20
2	Ice caps and glaciers	2.14
3	Groundwater	0.61
4	Surface water	0.009
5	Soil moisture	0.0005

Source: US Geological Survey (2018)

development report (UNESCO, 2024) discusses 'Water for prosperity and peace'. In dealing with the status of the world's water resources, the UN expects climate change to intensify and significantly impact the global water cycle, and to further increase the frequency and severity of droughts and floods.

7.2.2 ADB's Division of the Water Sector

Before proceeding, it is useful to present a summary of how one of the major development banks, the Asian Development Bank (ADB), approaches lending in the water sector (ADB, 2022). The sector is divided into three main sections to more accurately monitor what is being done now and in the future, so that progress can be properly evaluated. ADB has divided the water sector as follows:

- **Agriculture:** irrigation and drainage
- **Rural:** flood protection, sanitation, and water supply
- **Urban:** flood protection, sanitation, sewerage, water supply, and water-based natural resources management (includes industrial and municipal)

Geographically, the distribution of useable water across countries is not evenly divided. Many African and Asian countries as well as Middle East ones, small and large, are lacking freshwater as do some southern European ones. Canada, a large country in area, holds 20% of the world's freshwater but with a comparatively small population for its size (40 million approximately). Similarly, Lake Baikal in Siberia, the largest freshwater lake in the world, also holds 20% of the world's resources, again somewhat remote from where people reside.

Food and fibre production have been expanding steadily over the past six decades, in line with population increase, which accounts for the more than doubling of the water demand in agriculture, from about 1,400 km^3/year in 1960 to about 2,800 km^3/year at the end of 2023. By mid-century, with the further increases in population, the usage will probably be close to 4,000 km^3/year. This very significant increase in use by the agricultural sector creates a number of challenges, apart from the result of growing competition from cities, which will also be expanding. Demand will also be increasing in sectors such as energy and industry. The consequent main challenges will result from drawdown of underground water. There will also be problems of water quality because of polluted runoff resulting from fertiliser and pesticide use.

Feldman (2012) in recalling the magnitude of the changes forecast, as well as those already occurring, listed them as follows:

- Increased droughts in equatorial/temperate regions.
- Mountain glaciers are melting at faster rates.
- Earlier onset of spring with photosynthesis occurring longer during the autumn.
- Snow and ice cover is decreasing and melting earlier.
- Total annual snowfall is increasing in the far northern regions.
- More water availability in the moist tropics and in higher latitudes.

The distribution of water resources will likely alter, in some cases for the worse. Sana'a the capital of Yemen, is quoted as the first city likely to run out of drinking water, although years of internal conflict with the interference of larger, powerful neighbours involved have not helped. This raises another issue, that is, that conflicts across the world are an additional hindrence to people, particularly the poor, having ready access to drinking water.

Northern Pakistan, which was referred to earlier in Chapter 3, suffered severe flooding of the basin of the Indus River in 2010–2011, more severe than usual, possibly as a result of climate changes affecting the melting of the Himalayan glaciers, as climate scientists had been forecasting. More recently, at the end of October 2024, the Valencia region of Southern Spain was hit by disastrous flooding. It received the whole month's average rainfall in just 8 hours. The damage and loss of life were catastrophic.

The international development banks, such as the World Bank and ADB, will have to give much attention to the water sector in its several target areas throughout the rest of this century at least, much as the UN has warned in their 1983 report. UN Sustainable Development Goal (SDG) 6 is concerned with achieving satisfactory levels of water availability and sanitation for all by 2030. None of targets associated with SDG 6 were on track worldwide as of 2022 (Table 7.2). It is worth noting that the problems do not just exist in the developing world but some of them are found in many of the more developed countries as well.

7.2.3 Asia Generally

Water security remains at the top of the Asian agenda for the foreseeable future. ADB report that

> despite progress in addressing poverty, an unacceptably large number of people are still without access to safe drinking water and basic sanitation, and the situation is dire. As a result of the coronavirus disease (COVID-19) pandemic, some additional 75–80 million people in the Asian region were pushed into extreme poverty in 2020
>
> *(ADB, 2024a).*

According to the WHO and UNICEF, almost 500 million people in the Asian region still do not have access to at least basic water supplies and 1.14 billion people lack access to basic sanitation (ADB, 2020), and wastewater generated by cities is discharged into water bodies without treatment (ADB, 2016). About 40% of the rapid urban expansion

TABLE 7.2 Lack of progress in meeting SDG 6 targets (ADB, 2022, UN, 2025)

KEY ISSUE	*STATUS-SHORTFALL TO 2020 TARGET*	*INCREASE IN PROGRESS TO MEET 2030 TARGETS*
Safely managed freshwater	Lacked by 2.2 billion people	6 times
Effectively managed sanitation	Lacked by 3.5 billion people	5 times
Handwashing facilities	Lacked by 2.2 billion people	3 times

takes place in low-income, high-density settlements, challenging the abilities of city governments to provide basic services. With the significant increase in the world population in the past 60 years, with more still to come, this situation is not confined to Asia, and some places are worse, for example in the African continent. Changes in demography have placed fresh demands on the water sector, and climate changes will place even more.

The Asian region generally has to recognise the significance of the demographic changes taking place in several countries, for instance due to the aging population. In the water sector, this will result in implications for farm or small-holding size and related water management. This can be either through division of the land because of inheritance or due to opportunities for land consolidation and more efficient irrigation practices.

7.2.4 Africa

The sub-Saharan part of Africa is forecast to show the biggest population growth area in one continent over the remaining part of this century, approximately 2.5 billion (Figure 1.2). In terms of water management, this area of the world is arguably the one that needs the most attention. The World Bank reports that as of October 2022, about 226 million people in Eastern and Southern Africa alone did not have access to basic water services, and 381 million people lacked access to basic sanitation services. The situation is worse in rural areas than in cities. Addressing people's need for water and sanitation and continuing to make progress over the whole of Africa is key to improving people's lives and boost economic growth.

There is growing understanding from the relevant countries requiring assistance, as well as the African Development Bank, the WB. They will be expected to become increasingly involved in helping these countries develop their needed water infrastructure. One should not ignore the potential development support from China which has, over the past decade, made a determined effort to carry out business activities in Africa, largely with a view to accessing the valuable natural resources to be found there. With a changing world, the end of the century will see different countries taking the initiative. As African countries continue to rapidly urbanise and climate change impacts on rain patterns and water supplies, the need for innovative and efficient water production, storage, and delivery is stronger than ever. The major challenge is to ensure that governments, private sector, and households continue to invest in water and sanitation solutions.

7.2.5 Water and the USA

In the USA, water does not feature in the government debate, as there is no apparent shortage so far; however, there is an increasing risk from changing climate issues, for example, particularly, in states bordering or influenced by Caribbean cyclonic events. There has also been recently highlighted by severe drought conditions and the shortage of water to tackle the devastating Californian fires in the Los Angeles area (2025), that have ruined thousands of homes and livelihoods and caused some deaths. This may have been caused by increased extraction of underground water over the past few decades causing a drop in the groundwater table. Again, the global ongoing loss of underground water, noted in particular in north-west America, may cause water to become an important issue in some part of the country.

7.3 PRIVATISATION OF THE WATER SECTOR

7.3.1 Aims

The aim of private sector participation in the management of the water sector is that it should lead to improvements in the efficiency and service quality of water resources. It is claimed that it allows for increased investment from the private sector and eases the burden of water sector development and management away from the government budgets and the taxpayers (Wikipedia, 2025). Privatisation supporters also argued that the state-owned water utilities were not focussed enough on building infrastructure for water in a manner that would allow the economy to grow sustainably. For this reason, the economically and politically powerful, including development banks, were in the forefront of a shift towards making water a privately owned utility in many countries, both developed and developing.

7.3.2 The UK's Water Sector Privatisation

One of the best-known examples of water privatisation in the late 20th century was that undertaken in the UK under Margaret Thatcher. Since 1960, the population had doubled in England, but limited new water-related infrastructure had been put in place. The privatisation of the water sector in UK initially commenced in 1989. England was divided up into eight regions according to the main river that went through the region in question, and taking note of the size of population served. Scotland, Wales, and N. Ireland each had one public sector and regulatory body working to their relevant devolved governments. Each of the private sector organisations have overseen all water matters in their region for some 35 years. This has not been the overwhelming success that was promised.

In the recent past, it has been noticed by the public, particularly those with environmental concerns, an unacceptable rise in the level of pollution, particularly untreated sewage. The management of the water companies must take most of the blame for poor performance and lack of awareness for the last two decades or so when their sector was obviously deteriorating, even though significant dividends continued to be paid to investors. Investment for maintenance and very probably capital works had been and was required. Newspapers and environmentally aware politicians have increasingly been raising attention regularly to the unacceptable state of English rivers.

The levels of sewage discharged into English rivers, particularly the ones involved in freshwater supply, have greatly exceeded the amount decreed as allowable by Ofwat, the UK Environment Agency for water issues. Sewage levels reached 36% of water pollution and agricultural waste 40%. The bulk of the latter comes from slurry, the excrement from dairy and beef cattle and pigs which is spread on crop fields and then gets washed into rivers when it rains. In addition, there are chicken droppings and chemical fertilisers, and other less common pollutants, that are harmful when entering the rivers. The i-newspaper reported that it will cost 51 billion pounds sterling to just clean up all England's rivers (Heath, 2025), which means it will take the budgets of several years to resolve this problem.

Awareness of the discharge problem is getting to the point that the general public has started to show their dissatisfaction by organising street marches, and becoming increasingly annoyed with the government and the private sector water companies contracted to look after a given area of the country.

As well as the deterioration in the water in each region, there is a growing shortfall in freshwater such that there is a requirement for ten new reservoirs. Two, near Cambridge and in Lincolnshire, are under construction but will not be completed until the early 2030s while the remainder, which have yet to commence, will not be ready until 2040.

The overriding reason behind failings in water privatisation is that successive governments have not paid enough attention until comparatively recently of their own role in the water sector when attempting to protect the environment. They should also be aware of the belated need to insist on quite steady and regular investments. The growth in population should have suggested to water companies and politicians that the sewage treatment works would have needed to expand decades earlier, but they now lie well behind where it ought to be.

This example raises a central issue, whether the responsibility for caring for the interests of all citizens is better handled by government, the public sector, or by private sector companies. Water company management must take at least some of the blame for poor performance and lack of awareness over the last two decades of sector deterioration and lack of investment to cover additional or replacement infrastructure.

In summary, there is little doubt that there has been a lack of focus on this very important issue by all parties carrying some responsibility for the sector, government, regulatory body, or company management. The government takes the blame for not foreseeing the problem and providing adequate interest or support, including funding, when required. The company management just accepting the situation and self-rewarding their incompetence and inadequacies have justifiably taken 'stick' from the public. More realistic fines on the water companies are now being applied by the regulator, Ofwat. This will henceforward be an ongoing issue.

7.3.3 Privatising Water in a Developing Country

For contrast we can turn to another example, presented by Feldman (2012), this time to a South American country, Bolivia. In 1997, in order to comply with a WB loan, the country's congress voted to turn over control of its water utilities to two corporations – one French and the other American. Both imposed strict rules on urban residents' ability to collect rainwater from their roofs, imposed charges for drawing water from private wells, and increased service rates by nearly 200%.

The residents naturally protested, and the water companies argued that the levy was necessary to expand the services and overcome past government corruption, money that could have been used to improve treatment of infrastructure and water delivery. In the end, some 8 years later, when the international organisations withdrew, a Ministry of Water was established and was given the task of overseeing supply and providing universal access. In spite of reforms, provision of clean water was still judged as being inadequate because the country had not developed sufficiently to invest properly in reliable freshwater supply.

These are two different examples given above, one in a well-developed country, and the other a developing country. The problems are different in some ways but they both needed government attention. The one of the UK is likely to be solved in time, its issue, which is more of one of neglect rather than lack of resources, and much more needs to be done. In the Bolivian situation, it would seem the government ministry needed to appoint a special body with authority, for a time, set up, and working beside the government, possibly financed by the World Bank, and including a handful of technical experts in the water sector. The body should include training experts/academics because Bolivia, it is clear, lacked trained people. Matters will take time and resources, possibly two decades, to train up people in the various disciplines that will be called upon. It should be remembered that in the examples quoted, they have currently adequate supplies of freshwater, which is not always the situation.

7.3.4 Indonesia's Experience from 1970

One equatorial country with substantial rainfall that has changed its outlook over development in the past 50 years, so that it is now one of the top 20 countries in terms of GDP, is Indonesia. At the start of the 1970s, the government was in the hands of the military, who took over peacefully from the civilian government. Its people were generally poor at that time and average longevity was just 40 years. By the mid-1990s, Indonesia was developing steadily with a population that reached 200 million by the turn of the century, at which time the average longevity had risen to 70 years. There was a need, however, to focus attention on developing the water sector, the highways sector already being much improved.

About 1997, with a WB loan, the country privatised the water sector in Jakarta, for over 20 million people (the same period that Bolivia, tried privatising their water sector with WB assistance). Two international companies, one British and one French, were engaged alongside suitable Indonesian companies in partnership as part of the learning process. The capital city of Jakarta was expanding steadily (Younger et al., 2016), faster than the upgrading and replacement work required. At this time, it is probably fair to say that aspects of the work were all right. A development, which is lagging behind, as in too many other big world cities, is establishing a comprehensive sanitation system, and there is still much to do in many countries, including Indonesia.

7.4 DRIVERS OF CHANGE

7.4.1 Overview

ADB (2022) identified a number of key drivers of change in development. Some of these can be directly related to issues of water resource and water management and are discussed in this chapter:

- Climate uncertainty
- Pandemics (health)
- The Digital Revolution
- Agriculture in Transition (land ownership)

7.4.2 Climate Issues

Climate change increases the risks inherent in water management issues. This is particularly notable in Asia but will be more widespread in the future. ESCAP (2012) reported that Asia was the most disaster-prone area in the globe, 40% of all disasters and 84% of all people being affected. Costs incurred in the decade to 2020 from floods and droughts were assessed at $244 billion, and major low-lying coastal cities were especially vulnerable to storm surges, rising sea levels, and sometimes additionally due to land sinking from consolidation of soft ground.

It is a significant issue, for example, in the north of Jakarta where the country's main port, Tanjung Priok, is located. A 30 km sea wall or dyke has been designed to protect the port's infrastructure and city front from rising sea levels and ongoing settlement of unconsolidated soft ground, but the cost has so far delayed the start of construction.

7.4.3 Water and Health

ADB has reported that poor water quality and the increasing abstraction of water are putting pressure on the environment, and that water quality in Asia has deteriorated significantly with pollution increasing in half the major rivers for the last two decades and will continue (ADB, 2024b). A similar finding was reported above when discussing UK river systems (The Rivers Trust, 2025). This emphasised that rivers, despite being fundamental for providing the water that we drink and use to grow food, to creating nurturing spaces for wildlife and community, have for decades faced challenges from pollution, abstraction, and modification, and are under increasing pressure as our climate changes, and our water demands increase.

Poor water quality contaminates water sources, affects food production, and harms health, thus constraining development. The rapid drawdown of groundwater aquifers has led to inequities in access and land subsidence in some major cities, as well as an increase in saltwater intrusion in coastal areas, for instance in West Java around the city of Jakarta. Not all people have direct access to the main water supplies; a significant number draw water from the underground aquifer which can lead to land settlement from consolidation of surface sedimentary soils commensurate with drawdown of the aquifer. This does not provide a long-term solution and is a problem for several Asian coastal cities, for example Bangkok.

Poor water quality water is a health issue, as well as a water supply issue. The most common health risk associated with poor quality drinking water, which can cause diseases such as cholera and dysentery, is contamination of the drinking water from animal or human excreta and the microorganisms to be found in the faeces. Furthermore, countries such as Bangladesh and India are highlighted as having contaminants such as arsenic and fluoride found in the groundwater, which also lead to health problems like skin cancer.

7.4.4 The Digital Revolution

It is of interest that the water sector has begun to look at the Digital Revolution, in particular, how AI-based techniques (Kirwan, 2024) can be applied to optimise reservoir operations, monitor water quality, and monitor as well as predict water-related disasters, like in flood forecasting.

There is a shift from current maintenance systems on routine inspections and work schedules to a more productive approach making use of intelligent sensors. It is predicted that advances in AI models, sensors, and robotics will allow mobile repair bots to detect and fix leakages, helping to reduce the significant amount of non-revenue water. It is recognised that much knowledge will have to be gained and expertise applied. Most of the next decade or so will be taken up in acquiring the relevant skills. Before 2040, we should see the most advanced societies applying or starting to apply AI to some or all of their maintenance operations. It is still early days with AI.

The potential savings in water use by applying AI are very important as the century advances. This will compensate for the increasing demand for water due mostly to further increases in population, not to mention the greater stress on the sector that has been caused so far by the growth in population over the past few decades, as well as the commensurate rise in pollution. There is forecast to be, however, a gradual decline in the population over the last two decades of the century, as depicted earlier in Figure 1.2.

7.5 AGRICULTURE AND LAND OWNERSHIP ISSUES

The arguments affecting agriculture relate to climate change, environmental issues, and dietary changes, which are seen as the most important, a mix of both short-term and long-term risks, reflecting uncertainties, and all having an impact on the water sector.

The drawdown of underground water is arguably the most worrying trend, and more attention will be required to raise the efficiency in food production, for example in the System of Rice Intensification (SRI), a methodology for growing rice that aims to increase productivity and reduce environmental impact by optimising plant, soil, water, and nutrient management.

The question of land ownership and the right to use the water underground or surface water, whether in a lake, aquifer, or river has raised many legal problems in the past. From the ownership point of view, they are significant because they allow use to be made of the water as a product that can be marketed. Most water rights are based on property rights, riparian legal systems, in countries sometime touched by the English legal system, the USA and UK, as well as the English-speaking Commonwealth countries, including parts of India and Africa. There is a duty to such ownership wherein watercourses are to be kept clean and able to be used by upstream users in such a way that downstream users are unaffected. In addition, the riparian rights system does not give the absolute right to a landowner to a certain quantity of water.

In the west of the USA, after independence from Britain together with much of South and Central (Latin) America, which fell under Spanish influence, there is a left-over of strong feeling of injustice felt by the indigenous peoples. The indigenous population who had been dwelling on the land for centuries, resented, with some justification the European newcomers. The rights to water often became a contentious issue, in which the indigenous people usually lost any argument that ensued.

7.6 WATER RESOURCE POLLUTION AND SANITATION

7.6.1 Wastewater and Sanitation

There needs to be serious attention paid to wastewater and sanitation. In the 1960s, when the world population was 3 billion, the level of attention to sanitation was, in overall terms, weak and continued as such in the years following. It is certainly a priority action now for ADB (2022) across Asia and the Pacific. ADB have a large number of projects related to wastewater and sanitation, some of which are listed in Table 7.3, and show the diversity of the undertakings they are supporting.

It can be seen that ADB is heavily involved in the Asia-Pacific region on wastewater and multifarious projects undertaking a range of different issues depending on the most pressing need in the country concerned. The area covered by the projects is dominantly southern Asia with a population of approximately 3 billion. The work done or being done is wide-ranging. However, while there have been marked improvements to the quality of life for many people, it still requires a tremendous lot of work to be done, probably taking several decades, and the pace of development for a given country as well as a dependence on the level of political stability that the country can have. Secondly, it will require good education and training. There are good universities in

TABLE 7.3 List of some of ADB projects in Asia showing diversity of project scope

COUNTRY	*PROJECT*	*DETAIL*
Kiribati	Reducing waterborne disease	Climate-resilient water supply and improved hygiene practices for poor households and especially those headed by women. Data to provide insights into the benefits to women of all day access to safe water.
Laos	Training female water utility staff	A successful ongoing initiative.
Cambodia	Mobilising youth in water sector	150 rural youths trained to build good practices for water, sanitation, and hygiene.
India	Treated wastewater	Wastewater reuse and faecal sludge management for reducing freshwater demand, pollution loads, and health risks.
China	Water–energy nexus	Improving efficiency in water supply. Water and energy closely linked. Improving energy efficiency in water supply and strengthening economic instruments and market-based mechanisms for the allocation of water.
Asia/Pacific	Water sector governance challenges	Assess challenges. Limited implementation and uptake of policies because of capacity and funding gaps; limited use of economic instruments to manage water resources, insufficient data, and monitoring; limited uptake of integrity tools; and low stakeholder engagement.
Indonesia	Improving irrigation service delivery	Using results-based lending to refocus system operation on outcomes, adopting a combination of high-level technology; participatory processes focusing on asset management; and needs-based budgeting, institution strengthening and climate resilience.

India, Singapore, Malaysia, Thailand, Philippines, and Indonesia, for instance, which would be willing to support courses to assist in developing cohorts of staff and operatives to deal with the water sector.

7.6.2 Sewage Issues

Sewage, particularly in towns and cities, has not had nearly enough attention in the past. Although this has been recognised by ADB in their project portfolio, the amount of work still to be done will take several decades into next century to show positive change. The speed at which sanitation works are completed will depend on what has been done to date and a different mindset from that exhibited in the past.

Most Asian towns and cities have been developed at a time when sewage works were not considered as essential. Each home would have had an individual septic tank arrangement which was expected to be maintained from time to time. To change the system to put in a modern arrangement where the raw sewage is directed to a main sewage treatment works for the area in question would be a very difficult and expensive undertaking, near impossible in some cases from the density of construction in place, as governments have found.

7.7 UNDERGROUND WATER: A DIMINISHING RESOURCE

Recently, scientists have reported that so much water has been pumped out of the ground, causing a significant deficit in the level of groundwater, that the Earth has tilted on its axis by almost 800 mm to the east in less than two decades (Bawden, 2024). It was estimated that a net of 2,150 billion tons of groundwater was drawn up from beneath the surface of the Earth in the 17-year period from 1993 to 2010 – a study reported in the Journal, *Geophysical Research Letters.* By extrapolation. It is assumed that in the following 14 years from 2010 to 2024, the figure would be closer to 4,000 billion tons of net extraction. This would not necessarily be evenly spread, the drawdown of the underground water table being more marked in some regions than in others.

World water is mostly used for drinking, irrigation, and livestock; India and to an extent the USA. and China have drawn on groundwater as a source of freshwater. With the world population expected to grow by nearly 4 billion over the next 50 years before gradually declining by the end of the century to approximately 11 billion, it should be of concern that this factor may cause quite an upset in overall world climate stability with possible change in some of the regions.

One must be reminded that the world has exhibited ranges of temperature as the climate of some of the regions has changed over the centuries: however, there has never been a water crisis caused by the groundwater supply being so diminished. Water will become increasingly a main priority in the future and before the end of this century and requires as much attention as the climate. Fortunately, the skills and the solutions are available to address the various issues. However. the demand will probably exceed the ability, funds and trained personnel, to meet the very significant needs in the short term. It will take longer to undertake the necessary work, probably into next century.

7.8 SUMMARY AND CONCLUSIONS

Of the 196 countries in the world that are members of the UN, many of them are poorly governed, the institutional capacity of the water sector is weak, and this impacts particularly on all the diverse divisions of the sector and the complexities that are accordingly derived. The financing challenges to achieving change in the water sector are considerable and this makes investment in the water sector difficult and not well defined. This is especially for so for irrigation which involves a large areas of land, the use of which can change as agricultural markets alter. Additionally, the effects of climate change will be considerable, as climatic extremes raise uncertainty in conventional agricultural planning approaches.

The realisation that there will be problems with freshwater production and sewage has only been recently fully appreciated, and the extent of the issue has not been widely understood. There is a major problem in terms of outstanding treatment of pollution, plant to be built for sanitation works, provision of drinking water, and putting back adequate quantities of underground water. Extraction of this may have as big an impact on climate change as putting excess quantities of carbon emissions.

Application of AI has been recognised as something that could be very useful in terms of maintenance of water operations, for instance, for determining quickly the source of leaks. It could help to mitigate the problems and speed up the solutions that would arise as significant expansion of the many networks are required. Within a few decades, desalination plants may be required to provide freshwater to at least balance some of the shortfall of underground supplies of freshwater.

One possible way to monitor the world's water is similar to what was suggested for climate, that is divide the world into 46 regions, based on Koppen-Geiger (1976) and Beck et al. (2018). It must be admitted, however, the big gulf in sustainable development, between the well-off countries and those lower down the scale, would suggest that dividing the regions up, as suggested by Koppen–Geiger, will not be universally popular, and some countries would prefer to proceed on their own. Provided everyone participates, agrees that underground water is something that requires everyone's attention and all, at least a majority, are involved, then progress can be made. There are bound to be disputes arising and they should be settled by an influential body of independent persuasion aligned to the UN working with the WHO and possibly the World Bank. Disputes are expected to happen in the future.

Each region will have to concentrate on elements of the water paradigm, which is appropriate to their region. For instance, irrigation is very important to some equatorial countries, and much less or not to regions which have a temperate climate or colder. All countries should aspire to providing drinking water, and again that is unevenly divided on the planet. In addition, a long-term view needs to be taken, and some regions or countries will need to agree a given supply from a neighbour well-endowed with water.

From now and in the decades ahead, beginning soon, much greater care needs to be given to the water sector. Attention cannot wait. Practitioners will have to be aware of what is also happening to the water sector on an international scale and not just the one in which they have a particular responsibility (Younger, 2025).

REFERENCES

ADB (2016) *Asian Water Development Outlook 2016: Advancing Water Security in Asia and the Pacific*. Available at: https://www.adb.org/sites/default/files/publication/189411/awdo-2016.pdf

ADB (2020) *Asian Water Development Outlook 2020: Advancing Water Security in Asia and the Pacific*. Available at: https://www.adb.org/sites/default/files/publication/663931/awdo-2020.pdf

ADB (2022) *Strategy 2030. A Directional Guide to the Water Sector*. Asian Development Bank, Manila. Available at: https://www.adb.org/sites/default/files/institutional-document/855231/strategy-2030-water-sector-directional-guide.pdf

ADB (2024a) *Paving the Inclusive Path toward Water for All: Policies and Cases Supporting A Human Rights-Based Approach in Asia and the Pacific*. Asian Development Bank, Manila. Available at: https://www.adb.org/sites/default/files/publication/989246/paving-inclusive-path-water-all.pdf

ADB (2024b) *Innovative Finance Approaches for Addressing River Basin Pollution: Combating Aquatic Biodiversity Loss in Southeast Asia*. Asian Development Bank, Manila. Available at: https://www.adb.org/sites/default/files/publication/977796/innovative-finance-river-basin-solution.pdf

Attenborough, D. and Bitfield, C. (2025) *Ocean*. John Murray, London.

Bawden, T. (2024) Our planet is shifting on its axis – and humans are to blame. *Published in i-newspaper by* Environment Correspondent, December.

Beck, H. E., Zimmerman, N. E. and Wood, E. (2018) Present and future Koppen-Geiger climate classification maps at 1 km resolution. Scientific Data 5:180214.

ESCAP (2012) Reducing vulnerability and exposure to disasters. *The Asia-Pacific Disaster Report 2012*. The Economic and Social Commission for Asia and the Pacific, Bangkok. Available at: https://digitallibrary.un.org/record/4045530?v=pdf

Feldman, D. L. (2012) *Water*. Polity Press, Maiden, MD, p. 200.

Heath, L. (2025) Cleaning up all of England's rivers 'would cost £51 bn. *Published in i-newspaper by* Environment Correspondent, January.

Kirwan, C. G. (2024) *Smart Cities and Artificial Intelligence*. Elsevier, The Netherlands.

Koppen, W. (1936) Das geographische system der klimate. In *handbuch der Klimatologie*, Koppen, W. & Geiger, G. (eds.), Gerbruder Bontraeger, Germany. (In German).

Lane, A., Norton, M. R. and Ryan S. (2017) *Water Resources: A New Water Architecture*. John Wiley & Sons, Hoboken, NJ. (c/o Long Newnton, Gloucestershire).

Paxton, R. (2009) Loch Katrine water supply to Glasgow, Scotland, 1855–1903. In Great Rivers History, Rogers J. R. (ed.), Proceedings of the World Environmental and Water Resources Congress, pp 126–131. ASCE, 2009. ICE Scotland Museum Archives. Available at: https://ice-museum-scotland.hw.ac.uk/wp-content/uploads/loch_katrine.pdf

The Rivers Trust (2025) *Impact Report in 2024–2025*. Available at: https://theriverstrust.org/impact-report-2024UN (1987) *Our Common Future (The Brundtland Report)*. Report of the World Commission on Environment and Development. Available at: https://sustainabledevelopment.un.org/content/documents/5987our-common-future.pdf

UN (2012) *UN World Water Development Report: Managing Water under Uncertainty and Risk*. United Nations, New York. Available at: https://www.unwater.org/publications/un-world-water-development-report-2012

UN (2022) *Water and Sanitation*. United Nations. Available at: https://sdgs.un.org/topics/water-and-sanitation

UN, 2025 SDG 6.*Clean Water and Sanitation*. United Nations Statistics Division.Available at: https://unstats.un.org/sdgs/report/2025/goal-06/#:~:text=Slow%20progress%20for%20integrated%20water,per%20cent%20reported%20limited%20implementation.

UNESCO (2024) *Water for Prosperity and Peace*. United Nations, New York.

USGS (2018) *Where is Earth's Water?* United States Geological Survey. Available at: Where is Earth's Water? Available at: https://www.usgs.gov/water-science-school/science/where-earths-water

Wikipedia (2025) *Water Privatization*. Available at: https://en.wikipedia.org/wiki/Water_privatization

Younger, J. (2025) Briefing: The future of water: a looming crisis, attention required. *Proceedings of the Institution of Civil Engineers – Civil Engineering*, 178 (4). https://doi.org/10.1680/jcien.25.00175.

Younger, J. S., Parry, D. E., Lubis, H. A., McLernon, A. W., Wignall, D., Hasan, D. and Benton, G. B. (2016) Jakarta; the world's second largest city: Parts 1 & 2. *Proceedings of the Institution of Civil Engineers,* 168(4):253–261.

Transport 8

Jasper Cook, Scott Younger, and David Wignall

8.1 INTRODUCTION

At a macro-scale, transport is a fundamental driver of regional economic development and international trade. In contrast, at a micro-scale, it enables individuals and communities to rise out of poverty and overcome social exclusion, connecting goods to markets and linking rural areas and market towns to large cities and the global marketplace (UN, 2016). In short, moving people and freight efficiently over both short and long distances remains vital to development.

The aim of this chapter is to review some key aspects of transport within the context of sustainability as described by the UN (2016) as the "provision of services and infrastructure for the mobility of people and goods—advancing economic and social development to benefit todays and future generations—in a manner that is safe, affordable, accessible, efficient and resilient, while minimizing carbon and other emissions and environmental impacts". Sustainable transport is not itself a primary SDG but it provides key support to other goals, for example SDG 1 to alleviate poverty, SDG 2 to achieve zero hunger and ensure food security, SDG 3 to ensure health and well-being and SDG 4 to provide access to education (Cook et al., 2017). Achieving SDGs will depend on achieving advances in sustainable transport.

Transport may be considered as involving four principal modes: roads, shipping, rail and aviation. In terms of sustainability and carbon emissions, the Intergovernmental Panel on Climate Change (IPCC) report that direct greenhouse gas (GHG) emissions from the transport sector accounted for 23% of global energy-related CO_2 emissions (IPCC, 2023). Transport-related emissions in developing regions of the world have increased more rapidly than in Europe or North America, a trend that is likely to continue in coming decades. In this chapter, the focus is on sustainable development issues associated with road transport and shipping together with associated ports. The importance of rail and air transport is summarised with respect to their historical significance to sustainable development.

 DOI: 10.1201/9781003566274-8

8.2 BACKGROUND TO ROAD TRANSPORT

8.2.1 Key Issues

"Roads are the veins and arteries of the body politic, for through them flow the agricultural productions and the commercial supplies which are the life-blood of the state" (Gillespie, 1850).

After efficient international or national transportation, goods may arrive at a high-capacity freight station still requiring transportation to their final destination. This last leg of the supply chain is often less efficient, comprising a significant portion of the total cost to move goods. This is sometimes known as the "last mile" logistics challenge, which is of particular concern in the rural development environment. In this context, the provision of road networks features directly in two of the UN SDG targets (World Bank, 2019):

> SDG 9.1 Develop quality, reliable, sustainable, and resilient infrastructure, including regional and trans-border infrastructure, to support economic development and human well-being, with a focus on affordable and equitable access for all.
>
> SDG 11.2 By 2030, provide access to safe, affordable, accessible and sustainable transport systems for all, improving road safety, notably by expanding public transport, with special attention to the needs of those in vulnerable, situations, women, children, persons with disabilities and older persons.

There are currently many millions of kilometres of roadway networks across the world, from the simplest pathway or track to the most sophisticated trunk road with a thick surfacing of concrete or bitumen mix. The value of such connections between locations has been known and addressed for several millennia (Pryor, 2004). Recent development has been rapid. Taking the UK as an example of a developed country, the length of the road network in the last 300 years has expanded from 12,000km of unsealed indifferent quality roads to today's figure of 400,000 km of good-quality roads, highways and motorways.

Increasing attention had to be paid to roads towards the end of the Victorian Age with the advent of the motor car. Until the end of World War II (WWII) the number of vehicles was relatively small while developments were taking place on vehicles, engines and their fuel. The main effort in road building took place in the second half of the 20th century. Commercial traffic increased, and there was a need to extend the route network and upgrade normal trunk roads to motorways.

Most recently in the last few years, the use of electric vehicles (EVs) has become an increasing trend with the intention to make a significant contribution to carbon emission reduction and the impact of climate change. By the turn of the century, EVs will dominate road vehicles; Norway, for example, has 80% of its road vehicle fleet as EV. Design strategies are still being developed for "e-roadways" both in terms of being equipped to wirelessly charge compatible EVs on the move and take account of relatively heavier EV axle loading (BBC, 2024; Simpson and Clark, 2023).

In the sections that follow, general guidance is given as to the planning, design, construction and maintenance of sustainable road assets such that they can provide good service throughout their designated design life.

8.2.2 Road Transport Tasks

Road transport is required for a wide range of tasks, from provision of rural farm access to international high-speed safe mobility of people and goods (World Bank, 2023). On socio-economic, engineering and safety grounds, these different road tasks demand a wide range of differing road types (Table 8.1).

Detailed definitions of networks road assets will vary between countries. Figure 8.1 provides an example taken from a typical developing country road design manual (Cook and Petts, 2023a).

Despite the variation road tasks, there is a common thread in the approach to their sustainability, involving the following key principles:

- A fit-for-purpose design compatible with the local environments
- Appropriate quality of construction
- Adequate road maintenance and management

In practical delivery terms, sustainable road transport can be considered as comprising three main pillars as shown in Figure 8.2 (Cook and Petts, 2023b).

In brief:

1. Roads must be provided that are effectively planned, designed and constructed to be fit for purpose.

TABLE 8.1 Variations in road transport tasks

CLASS	*TYPE*	*PRIMARY TASK OR PURPOSE*
A	International expressway	Efficient mobility of people and goods on international standard highways.
B	National inter-urban highway	In-country transport of people and goods between and through major urban centres.
C	Rural feeder road	Linking local (provincial) capitals and other major urban centres with rural networks or other feeder roads.
D	Low-volume rural road	Provision of motorised access between small towns and villages, within local road networks and in agricultural areas.
E	Basic access	Provision of non-motorised and limited motorised access in rural areas. May not be within formal road authority jurisdiction.
U	Urban	Transport within major urban areas – which in itself may vary greatly in terms of traffic volume and vehicle type.
S	Special purpose	Roads with very specific tasks; e.g. quarry access, military access or tourist/religious purposes. Can have very variable axle loading requirements.

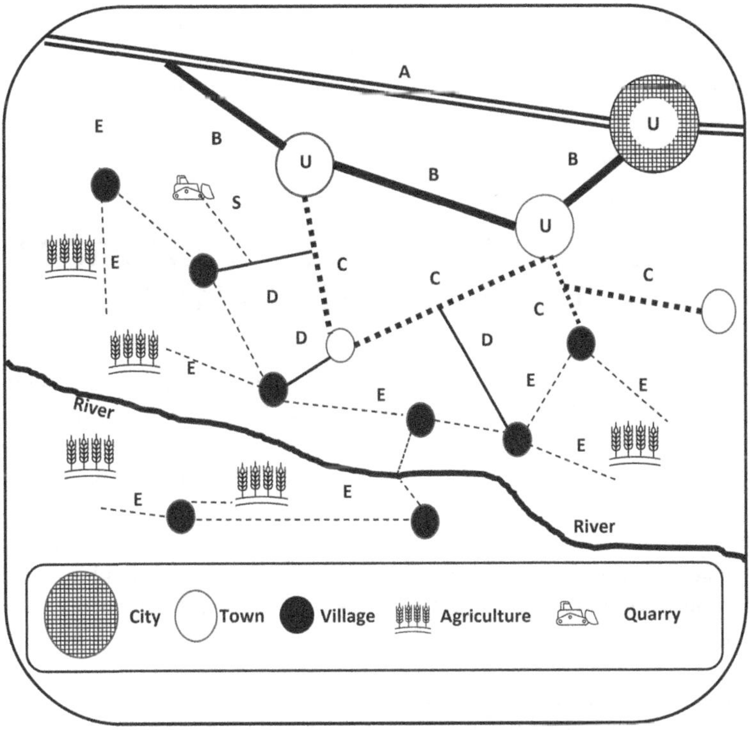

FIGURE 8.1 A schematic depiction of a typical developing country road network.

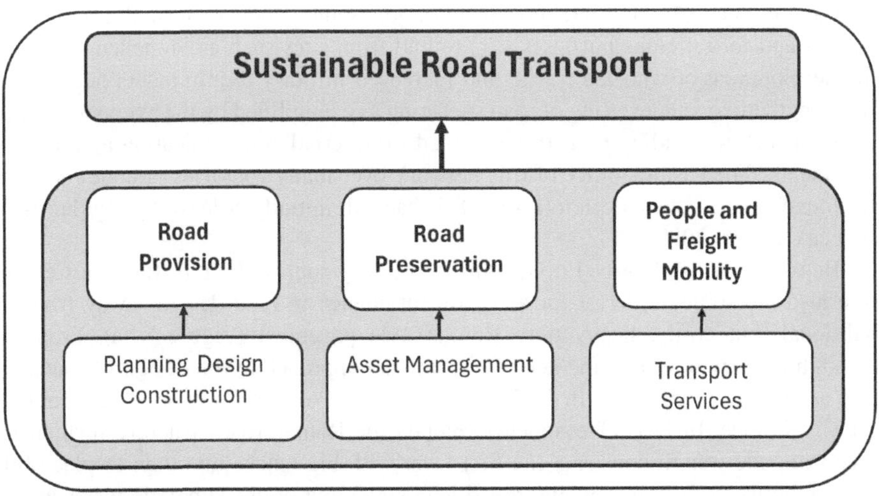

FIGURE 8.2 The sustainable transport continuum.

2. Road assets must be effectively managed in terms of regular and appropriate maintenance to preserve their fitness for purpose.
3. Safe and affordable transport services must be available to move people and freight along the provided roads in an eco-friendly manner as is possible.

8.3 SUSTAINABILITY LESSONS FROM HISTORICAL ROAD DEVELOPMENT

Engineered roads have a long history of development going back many thousands of years, even prior to the introduction of the wheel (in Mesopotamia around 3500 BCE). To be sustainable, even the earliest engineered access routes can be seen as responding to the same physical, climatic and socio-economic constraints as modern roads and from them we can identify lessons being learnt as to good engineering practice, sustainability and on the need to work within their defined environment (Table 8.2).

Some key drivers of change in road planning can be identified in historical of road assets and network and can be taken forward as lessons to be applied in sustainable development. These are listed and discussed below:

- The importance of strategic task
- Response to traffic and vehicle types
- Uptake of knowledge and research
- Importance of urbanisation
- Road management

Animal tracks were initially developed as largely un-engineered trails for basic agricultural foot-access to facilitate movement of goods and people to agricultural areas, to markets and for religious purposes (e.g. to and from sites such as Stonehenge in UK). The development of wheeled traffic and increased military requirements necessitated the strengthening and levelling of road surfacing as exemplified by the extensive Roman Road network responding to a strategic need for internal communication and military movements. The sustainability of this network over many centuries is undoubted and was founded on a robust design (Figure 8.3) backed, initially at least, by regular maintenance (Davies, 2011).

Between 17th and early 19th centuries, military supply chain traffic requirements were also the strategic driver for a significant change in road designs away from the traditional, but costly, heavy duty Roman-style pavement (Figure 8.4a). From the mid-18th century onwards, the need for a rational approach to the design of roads and their assets was recognised. In 1747, the "Ecole des Ponts et Chaussess" was founded in Paris, France. In 1765, Tresaguet developed the Roman road structure further. His basic principle was to construct the first layer with big blocks and then to place little rocks in between. In this way, improved load spreading on the subgrade was achieved (Figure 8.4b). These changes recognised that reliable and sustainable road networks

TABLE 8.2 Some key sustainability issues in an historical context

HISTORICAL ROAD ISSUE	*DESCRIPTION*	*HISTORICAL EXAMPLES*	*REFERENCES*
Alignment	Evidence of beneficially located alignment to suit improved access.	Ridgeway tracks in UK (4000–2000 BCE) moved away from soaked valley bottoms. "Sweet Track" wooden causeways lifted above weak swampy ground, Somerset, UK (2000–3000 BCE).	Lay (1992) and Pryor (2004)
Road structure	Roads constructed with a structure aimed at supporting traffic over natural ground insufficiently strong by itself to support anticipated traffic.	Earliest engineered road using defined layers of stone slabs, cobbles gravel. Heavy-duty Roman administrative or military roads. Development of more scientifically based stone road structures by Tresaguet in France, further modified by Telford in UK. Step change to lighter structure by McAdam with more task-based design with stone aggregate layers (Figure 8.3). Compatible with later use of bitumen and overlying concrete.	Lay (1992), Davies (2011) and Skempton et al. (2015)
Drainage	Provision of the means to remove potentially damaging water away from the roadway.	Roman roads provide with raised alignment and ditches. General Wade Military Road – adequate structural design, but poor drainage. Tresaguet, Telford, McAdam: Side ditches incorporated into design.	Lay (1992)
Construction materials	Roads constructed using locally available and appropriate materials.	Early use of brick (1500 BCE) in India. Different Roman Road designs in UK depending on location and geology/materials availability. Telford designs based on more easily manufactured aggregates. Use of bioengineering to stabilise weak surfaces, 2000 BCE (corduroy roads in UK and Europe).	Lay (1992), Davies (2011) and Pryor (2004)

(*Continued*)

TABLE 8.2 (*Continued*) Some key sustainability issues in an historical context

HISTORICAL ROAD ISSUE	*DESCRIPTION*	*HISTORICAL EXAMPLES*	*REFERENCES*
Provision of surfacing	An additional surfacing to aid the effective smooth and safe passage of pedestrian, animal or wheeled traffic.	Gravel wearing course on Roman roads. Gravel wearing course on Telford/McAdam designs. Introduction of coal tar, then bitumen waterproof seals from late 19th century.	Davies (2011), Lay (1922) and Towler and Dawson (2009)
Fit for purpose	Roads designed to suit a clearly defined transport purpose.	Roman military roads. Telford and McAdam design specifically in response to increased coach and wagon with potentially damaging metal tyres.	Davies (2011) and Lay (1992)
Maintenance	Proactive repair of roads to ensure their long-term service life.	Roman road shows evidence of regular repair of ruts before they broke through structural "base" into underlying "sub base".	Davies (2011)
Road funding	Accumulation or provision of funds specifically for the upkeep of defined roads.	Use of both central (Emperor) funds and local administrative funds (Province level) set aside for drainage. Gathering of tolls on "highways" to be used partially for upkeep of road condition (from 16/17th century in UK).	Davies (2011) and Lay (1992)

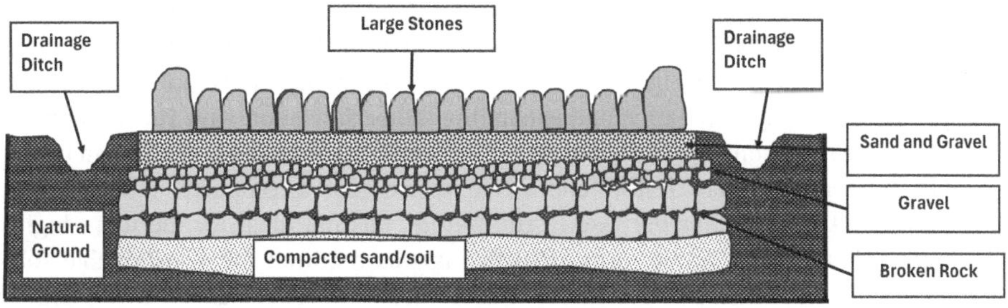

FIGURE 8.3 A typical Roman road cross section.

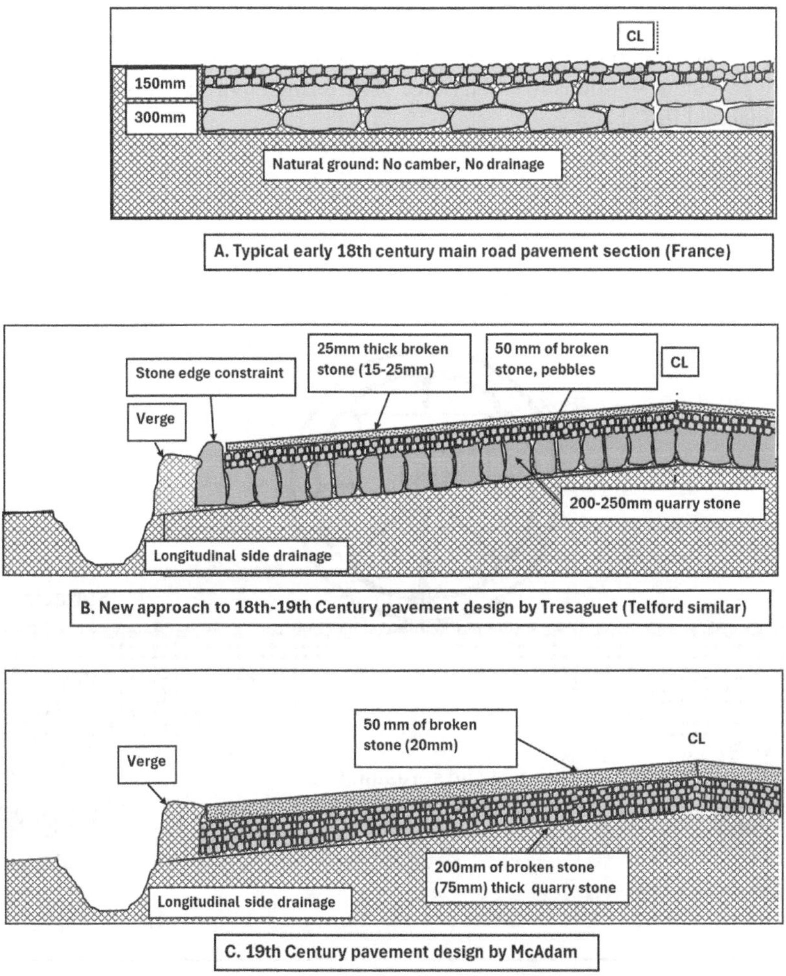

FIGURE 8.4 New approaches to road design 18th and 19th century (Europe).

had to take into account not only a strong pavement but also associated raised vertical alignment and drainage issues. These ideas were more fully developed in the designs by Telford and McAdam in the UK (Figure 8.4c). Following the Napoleonic Wars in Europe, an increasing population and expanding industrialisation gave further impetus to apply these improvements in road design in response to increased horse-drawn traffic.

The general designs of Telford and McAdam provided a compacted stone running surface that although fit for purpose in rural areas was not ideal for the more highly trafficked urban streets. Some form of surfacing was required for a sustainable solution. In the USA and Europe, from about 1880, tar from the local gasworks was sprayed over roads to make a dust and waterproof surfacing after being covered with gravel or crushed chips and compacted. From the 1920s, development of the automobile and increase in traffic volumes created a demand for improvements in the quality of road surfaces through the use of road grade bitumen (as a by-product of oil refinement).

From then on, road pavement engineers had sufficient knowledge of the basic load-bearing and running surface structure (Figure 8.5) required to provide sustainable pavements and to undertake research and development of the wide range of fit-for-purpose pavement solutions to suit a range of differing environments (Austroads, 2017) as summarised in Table 8.3.

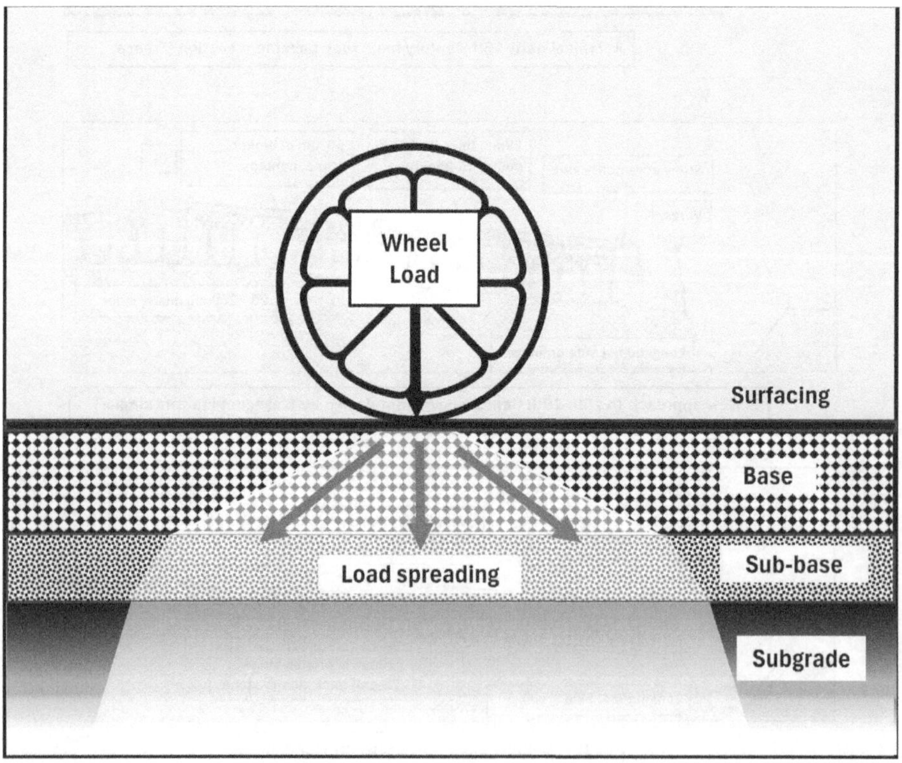

FIGURE 8.5 A fundamental pavement structure.

TABLE 8.3 Some typical pavement options

ROAD ELEMENT	*OPTIONS*
Bituminous hot seal Emulsion bituminous seal	Chip seals (various) Thin asphaltic concrete Sand seal
Load-bearing surfaces	Asphaltic concrete Hot premix/cold premix Reinforced concrete (various options) Reinforced concrete slabs Concrete blocks Cobble stone
Base/sub-base	Macadam Graded crushed stone Chemically stabilised stone/sand/soil Bitumen-stabilised stone/sand Mechanically stabilised sand/stone/soil

In the UK, the need for the transfer of military engineering knowledge to the civilian sector was facilitated by the founding in 1819 of the Institution of Civil Engineers. The application of road research into sustainable road design became increasingly important in the 20th century, and this trend continues (Parkman, 2014). In the UK, in 1933, the Government set up the Road Research Laboratory, which was upgraded to the Transport and Road Research Laboratory in 1972. In 1992, it was designated a government agency as the Transport Research Laboratory (TRL) and eventually privatised in 1993 as TRL Ltd. Many other countries followed a similar path in developing their research into sustainable roads; for example, Australian Road Research Board (ARRB), Australia; Transport Research Board (USA); and, on a wider scale, the World Road Association (PIARC). With the support of the British government, TRL has exported its expertise to other countries across the world, principally to Africa and Asia, where some countries established road centres on the TRL model (e.g. Malaysia, Myanmar, Kenya, Bangladesh). TRL Road Notes and Overseas Road Notes have become and remain a primary source of sustainable road knowledge.

In Chapter 5, a distinction was drawn between the rural and urban developments from an environmental viewpoint. There is a similar distinction to draw between the transport requirements of the urban populations compared to non-urban population. Today, the urban divide is calculated to be about 55% of the population, or about 4.5 billion; however, the world is becoming increasingly urbanised and 68% of the world's population is projected to live in urban areas by 2050. Modern industrial economies, economic and social opportunity depend on access to urban jobs and services (World Bank, 2019).

Improving urban transport efficiency has its own specific challenges that are increasingly being considered in terms of a shift in strategic planning focus away from purely traffic-oriented planning, with a primary goal to improve vehicle travel conditions towards mobility-oriented planning. This latter focus considers planning more in terms of the movement of people and goods, and the increased use of public transit, bicycles and even pedestrianisation. This will have a significant impact on the design of roads (or streets) in terms of, for example, cross-sectional geometry, task separation, running surfaces and foundation strength.

Road maintenance plays a key role in the sustainable road transport continuum from the planning of access provision through to effective transport of people and freight. The performance of maintenance management systems is dependent on a combination of interdependent factors including internal institutional arrangements, sustainable funding and technical capacity.

A maintenance sustainability pyramid concept (ReCAP, 2019) has been developed over a number of years that includes committed, adequate and sustainable funding for road maintenance as a vital element. This assumes an annual valuation carried out of road infrastructure assets, a costing framework and a budgeting and programming processes are in place for prioritised maintenance plans. This block requires the existence of committed, adequate and sustainable funding for road maintenance. Without adequate road network maintenance and funding, there is little chance for long-term transport sustainability goals being achieved.

8.4 THE ROAD ENVIRONMENT

8.4.1 The Road Environment Concept

Lessons learnt from the historical development of roads indicate their design-life performance, and sustainability is influenced significantly by complex interactions within a "Road Environment", comprising engineering, operational and specific sustainability criteria (Cook et al., 2013). It follows that these factors should be crucial elements underpinning the road design and management processes. A full understanding of the influence of the factors is essential if the roads to be constructed and maintained are to be fit for purpose, sustainable and provide value for money (see Figure 8.6) (Cook and Petts, 2023c).

8.4.2 The Engineering Environment

This environment is defined by a group of factors directly related to engineering issues to be addressed in the design process.

Climate: All types of roads are vulnerable to climate hazards and their consequences. Both current and future climate will impact upon the road in terms of direct erosion, flooding and the influences of the water-table changes on earthwork stability and hydraulic design criteria (Cook, 2022). Many large cities, in particular, are increasingly at risk from the consequences of sea-level rise (Alcaraz and Rigaud, 2023).

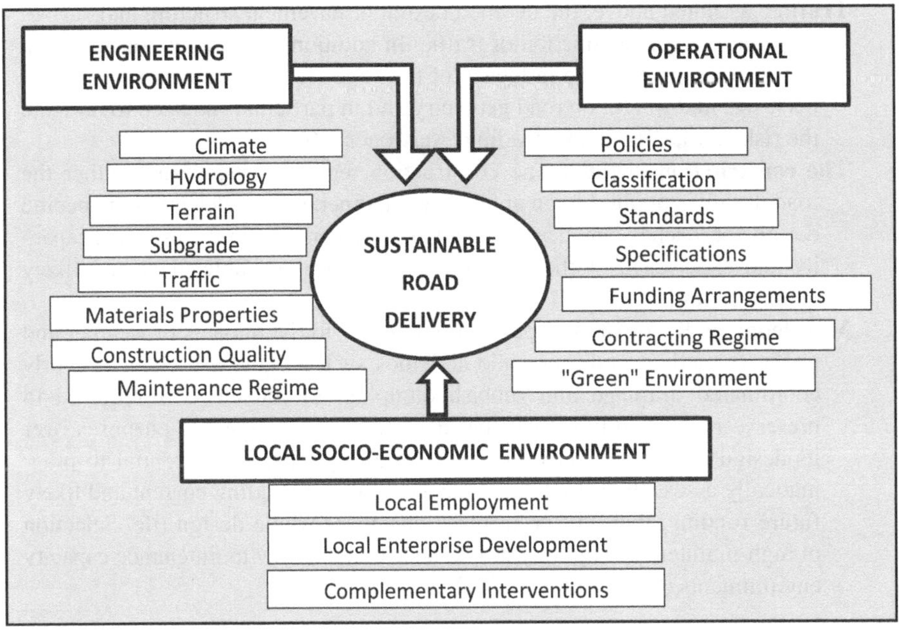

FIGURE 8.6 The road environment.

Hydrology: Water is the single most important factor affecting performance of rural road pavements, earthworks, drainage and long-term maintenance costs. There is a direct relationship between current and future climates and the assessment of hydrological factors which are a vital part of the hydraulic design of drainage, bridges and culverts (ADB, 2020).

Terrain: It reflects geological and geomorphological history and has a direct influence on both horizontal and vertical road alignments and associated earthworks. It is a key issue in the application of geometric and safety standards and the amount by which they may need to be adjusted for specific locations or specific road tasks.

Construction material properties: The nature, engineering character and location of construction materials are fundamental aspects of the road investigation and design. Sustainability as a whole requires that greater attention be paid to the use of environmentally friendly and re-usable materials (Austroads, 2020). The road sector is no exception to the need to minimise use of scarce primary resources and reduce carbon emissions. This is becoming ever more urgent as humanity steadily exhausts the planet's natural resources and carelessly discards waste to the detriment of the environment (Sasidharan et al., 2019).

Subgrade and foundations: A sound and fit-for-purpose pavement foundation has historically been recognised as an essential element in a sustainable road (TRL, 2023). The subgrade is the foundation layer for the pavement, and the combination of subgrade and traffic is an established relationship fundamental to the structural design of pavements.

Traffic: As noted above, the detailed design of pavement structure and surfacing type is a direct function of traffic. In addition, significant consideration needs to be given to the influence of traffic in terms of vehicle types, numbers, size and weight on road geometry and in particular the need to mitigate the risk of wheel/axle overloading (Austroads, 2017).

The construction regime: The construction regime can govern whether the road design is applied in an appropriate manner, or not. Selection of specific design approaches may have to be adjusted depending on the assessed capacity and experience of the available contractors, available plant and likely quality control.

Maintenance: Roads of all types deteriorate due to the impacts of weather and traffic as well as anthropogenic activities such as land-use change, poorly coordinated drainage and rubbish dumping. Maintenance is required to preserve road condition such that it is fit for its designated purposes over its design life. When selecting road design options, it is essential to pragmatically assess the actual maintenance regime, including current and likely future funding, that will be in place during its whole design life. Selection of high-maintenance options should be avoided in low-maintenance capacity environments (Cook and Petts, 2023d).

8.4.3 The Operational Environment

This environment comprises a group of factors that define the framework within which roads have to be planned, constructed and operated and hence should be a central basis for their sustainable design.

Policies: National or sectorial policies will provide guidelines, requirements and priorities for the road network decision-making processes. They provide the framework within which the road design must be undertaken. Policy frequently drives the strategic guidelines on which appropriate fit-for-purpose road design is based. For example, recent work by World Bank and UN (SuM4All, 2019) has highlighted the changing emphasis in transport planning of the urban environment on green public transport alongside, bicycle and pedestrian traffic and a decrease in private vehicle traffic. Such policy changes should have direct impact on the geometric and structural design of urban road networks.

Classification: Road classification essentially guides the road designer towards appropriate design and construction standards. Road classifications based on task or function provide road planners and designers with a practical guidance framework to initially select and cost appropriate road options.

Road design standards: Design standards provide the link between the cost of building the road and the benefits to road users. From the road designer's point of view, standards governing issues such road geometry, structures, pavement, safety or drainage govern a minimum level that must be achieved by a design if it is to be considered a sustainable solution. Additional standards govern the whole range of investigation, construction and maintenance actions.

Technical specifications: Technical specifications, appropriate to the local engineering environment, define and provide guidance on the design and construction criteria for roads to meet their required level of service within the governing matrix of standards and classification. Road designs should be either in agreement with the local technical specifications or compliant with approved modifications.

Funding arrangements: Available funding has an overarching influence on the scale and nature of the roads that are feasible. The assessment of funding as a constraint on road design should include not only available budgets for design and construction but also those available for ongoing road management and maintenance.

The contracting regime: The nature of the general contracting regime for either construction or maintenance influences the design and construction of a road through issues such as local legislation and contract documentation, contracting model, national or international contractor requirement, or use of local community groups or **small medium enterprise**s for construction of rural roads.

The green environment: Road construction and ongoing road use and maintenance have an impact on the natural environment, for instance for flora, fauna, hydrology, slope stability, health and safety. These impacts must be assessed, and adverse effects mitigated as much as possible by appropriate design and construction procedures (ADB, 2003).

8.4.4 The Sustainability Environment

Experience from within the development sector has shown that the sustainability of infrastructure initiatives, including roads, is particularly dependent upon being compatible with important issues specific to sustainability (Cook and Petts, 2023e), as shown in Figure 8.7.

8.4.5 Policy and Political Support

Road networks, and their component road links, must be compatible with identified national policy or development initiatives and have the full support of all relevant Ministries or Departments at central as well as at local levels. The central government support acts as the top-down driver and facilitator by providing the necessary strategic guidance for financial and technical issues as well as appropriate institutional, legal and regulatory frameworks.

Experience has shown that, to be fully sustainable, road initiatives require buy-in and support from relevant urban, local provincial, district, communities and groups. In addition to supplying local knowledge, a sense of ownership can add significantly

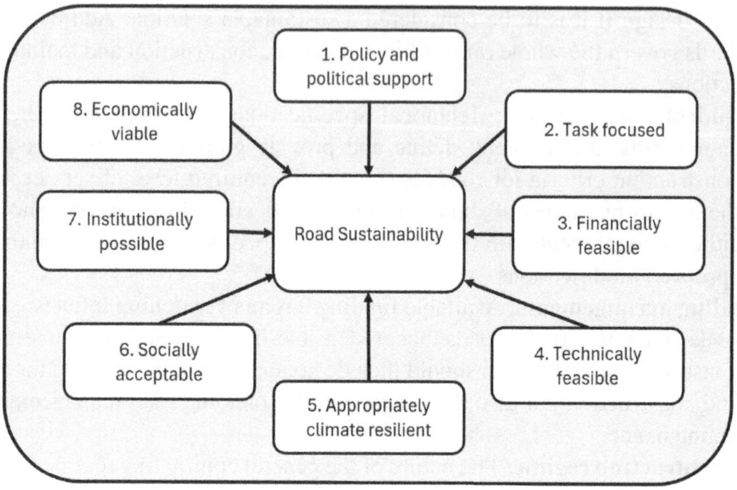

FIGURE 8.7 Sustainability criteria.

to bottom-up local focus on appropriate design and support to important sustainability issues such as routine and periodic maintenance.

8.4.6 Road Task or Fitness for Purpose

Fitness for purpose as has been noted previously, is the overarching driver of appropriate road design since earliest times. Roads must deliver a level of service in line with their identified purpose or task, which includes compatibility with the nature of the traffic (people and freight) which will pass along them, whilst minimising costs over the whole life of the road.

8.4.7 Financial Feasibility

Without adequate funding or funding mechanisms in place, sustainable design, construction, management and long-term maintenance of the road and its assets are not possible. The financial feasibility of a road project must be assured in terms of the total expenditure incurred over its whole design life.

8.4.8 Appropriately Climate-Resilient

Transport infrastructure is known to be vulnerable to current and future climate threats and associated impacts. Roads are built to provide sustainable access, and if this is compromised by climate impacts, then this can have significant economic, social and financial consequences and can threaten the hard-won transport development achieved in the past few decades. Additional appropriate climate strengthening measures may be required (ADB, 2014).

8.4.9 Technical Feasibility

Road designs are required to comply with national classifications, standards and specifications that reflect socio-economic and transport needs. Design options should be constructible and maintainable using appropriate resources, and maintainable within the existing road regime and resource base. Designers should be wary of investigation, design or construction methods that have been found to have been successful in other road environments but are yet unproven in the environment in question.

8.4.10 Social Acceptance

Roads should be positively accepted by national and local stakeholders as being required initiatives. Road construction should comply with social safeguards on issues such as 1. improved road safety; 2. follow-up safety training, especially for vulnerable road users; 3. resettlement and land acquisition; 4. gender equality and 5. protection of vulnerable groups. In addition, what are termed "complementary interventions" can significantly enhance the social acceptability aspects as seen from a local level.

8.4.11 Institutionally Appropriate

To be sustainable, a road project should be within the technical and managerial capacity of the appropriate road authorities to carry it forward throughout its life cycle from design to construction and maintenance. Essentially, there needs to be a resource pool of road designers, contractors and road managers available with the necessary appropriate resources, knowledge and experience to implement a proposed sustainable road programme. This is an important issue, particular for rural or remote road networks, when considering the implementation of innovative design or construction possibilities, such, for example, as those associated with climate resilience or marginal construction materials.

8.4.12 Economic Feasibility

At its simplest, the assessed benefits accruing from the project in terms of socio-economic development benefits (and their assessed value) must be greater than its initial and ongoing costs or the assessed consequences of a do-nothing option. Often the appraisal will include an analysis of the net contribution that the investment in a sustainable road will make to society.

8.5 GOOD PRACTICE IN THE WHOLE-LIFE ROAD CYCLE

8.5.1 The Road Cycle

Most of the features of a road project can be related to the need to take a series of decisions and actions linked to its logical progression from inception to delivery of design life performance. A whole-life road cycle (WLRC) as shown in Figure 8.8 can be adopted as a framework around which road asset and road network management can be structured.

The WLRC can act as a route map to the relevant data and the appropriate levels of information and procedures required for making knowledge-based decisions. The use of a life cycle model also allows practitioners flexibility in terms of entry point, as many road initiatives are unlikely to start with new alignments, with most being based on the upgrading of existing roads or urban streets. Hence, a more appropriate entry point could be at Stage 6 (Rehabilitation and Upgrade Assessment) rather than Stages 1 or 2.

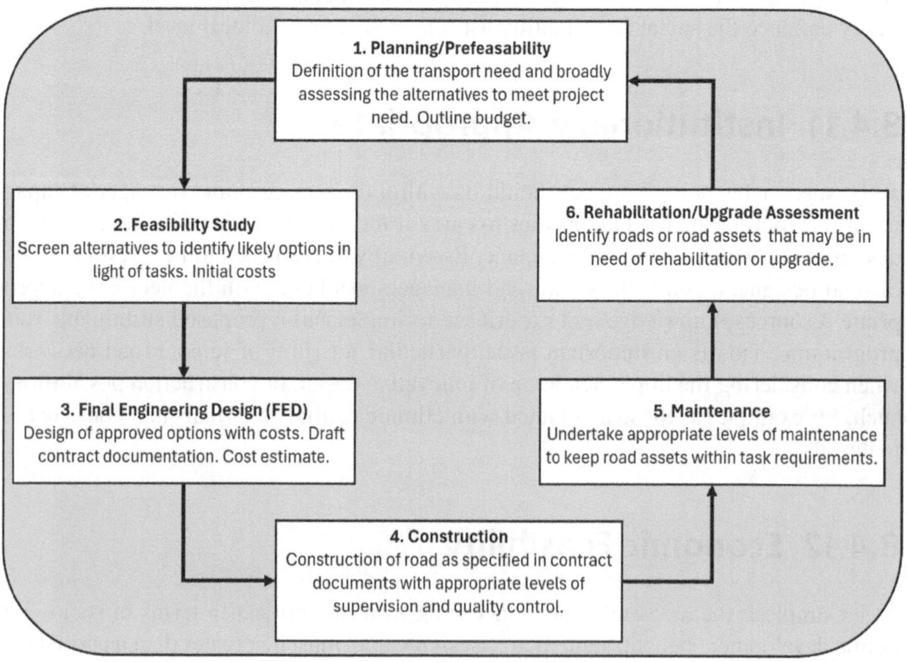

FIGURE 8.8 The whole-life road cycle.

8.5.2 Climate Resilience and Deterioration

As noted previously, all levels of road type are vulnerable to climate impacts. Current and future climate impacts work with other issues to place roads and their assets at risk of unsustainable premature deterioration. Key issues within a climate-related road deterioration matrix are as follows:

- Current climate
- Incremental climate change
- Shock climate events
- Land use
- Design
- Poor construction
- Lack of maintenance
- Axle overload (pavements only)
- Ineffective drainage

A logical accumulation and assessment of relevant knowledge can be cost-effectively within the WLRC.

8.5.3 Good Practice in Planning and Feasibility Studies

The early phases of a road project are where the foundation for sustainability is underpinned by the adoption of a good practice in key areas, such as the following:

Desk study: Collate available information on policy, masterplans and project background. Identify general physical and climate setting and sources of more detailed information.
Road task: Clearly define the purpose of the project within any specific transport policy and strategy and identify the appropriate classification and general standards.
Site visits: Identify key features and potential meetings with all stakeholders. Identify areas of potential hazard and climate vulnerability.
Initial field investigations: Identify and define key geotechnical and materials issues, traffic counts and axle loads assessments.
Construction materials: Collate local materials resources data particularly with respect to reusable/environmentally friendly materials.
Local technical resources: Collate data on relevant resources and skills capacity.
Climate: Define the broad regional climate characteristics and identify general levels of climate hazard.
Major structures: Locate any major and potentially costly structure.
Maintenance regime: Review current asset management capability.
High-risk issues: Early definition of major high-risk or high-cost challenges.

8.5.4 Good Practice Design and Construction

Areas where good practice plays a key role in sustainable design and constructions are as follows:

Detailed ground investigations: Following recommendations from a Feasibility Study in terms of materials sources, alignment and structures.
Local contracting and skills capacity: Select technology options compatible with local capacities and identify any enhancement needs.
Final engineering designs: Detailed phase of pavement designs; embankment and cut-slope designs, drainage and structures.
Climate resilience applications: Ensure future climate criteria are used for detailed design of earthworks, vertical alignment and hydrology input to bridge and culvert designs.
Asset management: Prepare detailed maintenance requirements and draft proposals for any required enhancement of funding or capacity.
Procurement documentation: Appropriate technical specifications, Terms of Reference, Bill of Quantity and general contract documents appropriate to contract model.

8.5.5 Good Practice in Road Maintenance

Good practice in road maintenance is key to ensuring a road will perform to its required level of service throughout its planned design life. Road maintenance is fundamental to the sustainability of a road network. Despite this, it continues to be poorly delivered, particularly at local road levels. Particular attention needs to be given to the funding levels needed to keep the road network in a satisfactory long-term condition (ReCAP, 2019).

Similar to the Road Environment concept, the Maintenance Environment (Figure 8.9) concept can be applied to road networks to provide a logical basis for developing and applying relevant maintenance models. Typical maintenance models that could be considered are as follows:

- In-house ("force account" or "direct labour")
- Conventional quantity-based contracting
- Performance-based contracting
- Large contractors (mainly main or provincial road)
- Community or village groups (mainly local roads)
- Small local contractor or SME
- Single-person contracts

The network environment key issues

1. **Road classification:** Size (width) of roads and the responsible authority.
2. **Road design:** Will define maintenance processes and required resources. Inappropriate design can impose severe strain on a maintenance model.

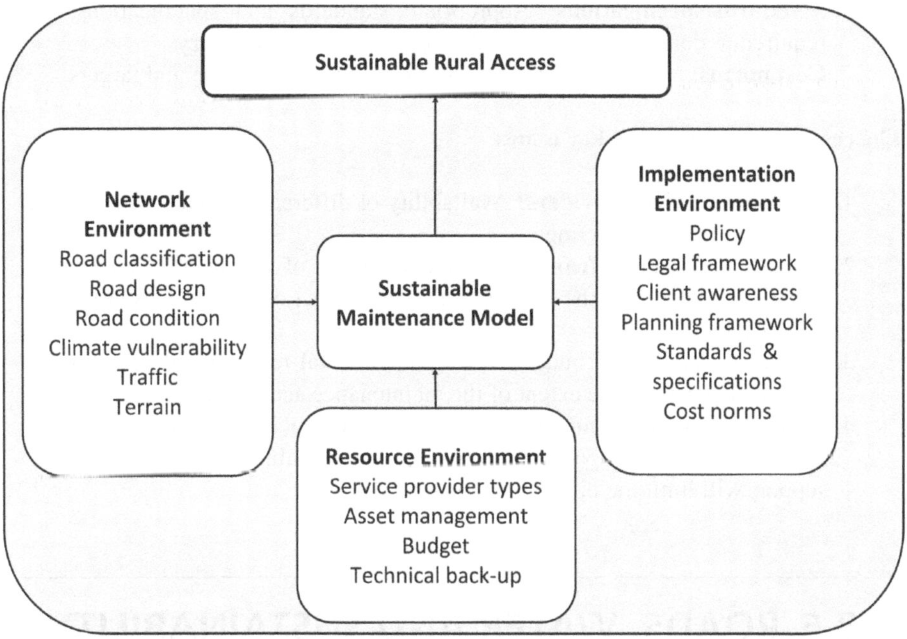

FIGURE 8.9 Maintenance environments.

3. **Road condition:** Poor initial road condition may imply the need for initial rehabilitation.
4. **Climate vulnerability:** Highly climate vulnerable roads imply a focus required on drainage and erosion protection maintenance and influence risks for service providers.
5. **Traffic:** Influences rate of surface deterioration, with a particular challenge in a high axle load environment.
6. **Terrain:** Linked with climate implies a focus on side drainage and surface deterioration.

The implementation environment key issues

1. **Policy:** The nature and clarity of the overarching policy on road network asset management will influence the effectiveness of the maintenance procedures and their application.
2. **Client awareness:** Experience within the Client Authority in the selected mode of maintenance is essential.
3. **Legal framework:** The in-place legal instruments can either enhance or limit the adoption of a maintenance model.
4. **Service provider capacity:** Appropriate service provider experience is essential.
5. **Planning framework:** An effective medium to long-term planning framework enhances sustainability in maintenance programmes.

6. **Standards/specifications:** Appropriate standards and specifications are required to define the activities and their service level delivery.
7. **Cost norms:** Need to be aligned with the maintenance mode and targets.

The resource environment key issues

1. **Available service providers:** Availability of different providers will govern the selection of contracting.
2. **Asset management framework:** An effective of the asset management framework enhances the options for model selection and their subsequent effectiveness.
3. **Budgets:** Appropriate budgets are a fundamental requirement, and limitations will constrain the extent of the maintenance activities.
4. **Technical support and QA:** The level of technical support and the QA arrangements will impact on the quality of the maintenance delivery. Poor support will limit the options for some models.

8.6 ROADS, WATER AND SUSTAINABILITY

8.6.1 General Principles

Water has been commonly described as the "enemy" of roads, through being potentially responsible for significant damage to this asset. However, the application of Roads for Water principles can also place roads in a development supporting role in terms of water harvesting and supply, working alongside overall transport network climate resilience.

Roads can be used to harvest and redistribute runoff to areas where it can be used. Road earthworks can be utilised to guide and channel rainwater to enable a road to systematically act as a water-harvesting mechanism. For this, the road-drainage mechanism needs to be appropriately designed. An effective combination of water harvesting and resilience adaption result in the following:

- Water being beneficially managed for the benefit of roadside water users
- The reduction of water-related damage to roads and thus reducing disaster risks
- The mitigation of adverse impacts of roads on the surrounding route corridor environment

Recent guidelines (World Bank, 2021) advocate a more integrated design process involving local communities and social, environmental, water and agriculture sectors to turn around the negative impacts of roads on the surrounding landscape and simultaneously maximise the beneficial use of roads for water management and climate resilience.

Water harvesting is perhaps most applicable in arid and semi-arid areas where it can be used to serve the communities in and around the road corridor (Van Steenbergen et al., 2018). The water intercepted by road assets can be guided to surface or subsurface storage or can be applied directly to the land. Given the significant lengths of roads being built, roads present a good opportunity for water harvesting and water management in many semi-arid areas (Salman et al., 2016). The principles of road-water harvesting are primarily applicable to low-volume roads where pollution and contamination from large volumes of passing motorised vehicles are not key issues.

8.6.2 Water Harvesting Application

The World Bank Guideline (2021) advocates an approach that is termed "resilience plus", involving integrating water management into climate resilient road design. This approach designs roads that are fit for purpose within the landscape, allowing them to support improved management of water and the local environment for the benefit of nearby communities.

The aim of the approach is to design roads and their engineering assets to collect runoff and guide it to recharge areas (Table 8.4).

TABLE 8.4 Road assets and water harvesting design

ROAD ASSET	*HARVESTING DESIGN FOCUS*
Bridges	Use bridge sills for controlled drainage and water management; consider drifts instead of bridges in appropriate locations.
Drifts	Use non-culvert drifts for water retention, river stabilisation and floodwater spreading.
Pavements	Amended alignment, geometry and cross drainage for water storage and recharge. Include basic drainage for water harvesting as part of road development; take measures to manage subsurface flows.
Roadside slopes/earthworks	Utilise bioengineering to stabilise slope faces and work with surface drainage to control and guide water to recharge areas.
Drainage structures	Surface and subsurface drainage to be linked to recharge areas. Place culverts to optimise drainage pattern for water harvesting. The road-drainage system is a mechanism for effectively collecting and diverting all water that is gathered around the road. If adequately designed, it can help "harvest" a large part of the runoff from the catchment uphill of the road and avoid waterlogging upstream of the road.
Borrow pits	Systematically convert borrow pits for storage, seepage or recharge. Plan new borrow pits to optimise storage functions after use.

Additional points to note with regard to this table are as follows:

1. Borrow pits may be reused for permanent water storage rather than backfilled, which is often done with low-quality soil material. This is a considerable cost-saving measure, and it creates a local water resource almost essentially free, although site-selection and safety measures must be considered in their planning and long-term maintenance.
2. Culvert-less, "non-vented" drifts, as well as preventing the scouring of rivers, encourage the buildup of sand-water storage immediately upstream of the drift, effectively combining their road access function with that of a sand-water storage dam (Neal, 2012).
3. Road embankments are reported as having been used for water storage, as in countries as diverse as Burkina Faso, Portugal, Turkmenistan, Uganda and the Republic of Yemen (World Bank, 2021). Because road embankment costs have already been included, reservoirs can be developed with comparatively minimal additional expense.
4. Road-drainage water should not be disposed of in areas where it creates damage or where it serves no useful purpose; rather, it should be led to agricultural areas, tree plantings, recharge areas or ponds. It is also important that road-drainage water (including the sediment and other particles it carries) not be directly discharged into streams where it will create turbidity and cause stream sedimentation.

The storage capacity of surface reservoirs such as ponds and borrow pits is limited to the size of the basin, but the water is readily available at the surface. Surface storage systems also suffer losses from evaporation.

In contrast, much greater quantities of water can be stored in the soil, in shallow aquifers, or both, provided that the geology of the area is suitable. The water storage capacity of soil varies according to the soil texture. By storing water in aquifers, evaporation losses are reduced compared with surface water storage.

World Bank (2021) considers that a large proportion of the runoff in a catchment can be retained by implementing several water conservation techniques, with intensive water harvesting potentially exceeding 30% of the runoff. Harvesting larger amounts of water can also reduce the volume of potentially destructive stormwater in the lower part of the watershed.

Reducing the speed of runoff also enables water infiltration to increase. Installing check dams, guiding water from steep slopes, building terraces and furrow ditches, and other techniques can be used to slow water runoff speeds. As more stormwater infiltrates the soil, less water must be managed as surface runoff.

As a general principle, the sustainability of road-water harvesting structures needs to be ensured and that includes their adequate maintenance and repair. Water harvesting structures may need regular inspection after each rainy season to address impacts such as erosion, and remediate any health, safety and environmental concerns. Water harvesting systems must include adequate measures to prevent mosquito breeding and waterborne disease arising from standing water.

8.7 ROAD TRANSPORT SUMMARY

In order to be sustainable, the following should be done:

1. Roads must be designed and constructed to be fit for purpose, within their physical environment, to meet defined tasks using sustainable and eco-friendly materials as much as practicable.
2. Roads must meet key specific criteria that are clearly focussed on defined sustainability needs
3. Roads must be managed in terms of regular and appropriate maintenance, protected against the impacts of excessive axle loading and upgraded where required to meet threats from future climate.

The addition of water-harvesting principles into rural road designs can not only add to the overall sustainability of the road itself but also make significant contribution to the overall sustainability of the road corridor and its agricultural development, particularly in arid and semi-arid areas.

8.8 MARITIME TRANSPORT

8.8.1 Historical Development

Transport by ship across the world on the oceans has grown greatly in the past few decades. While passenger numbers declined because of the ready availability of faster, cheaper, air travel, this has been more than balanced by increased bulk and container cargo transport and the rise in cruise ship traffic. Cargo tonnage moved today is more than 11 billion tonnes and expands at the rate of approximately 3.2% annually. In particular, this is the age of the large container ships with a gross tonnage in the order of 200,000 dwt. The largest of which is close to 240,000 dwt, carrying 24,000 20-foot containers (Twenty-Foot Equivalent).

From earliest times, maritime transport has been closely intertwined with human development in terms of migration, exploration and trade, as well as, unfortunately and inevitably, war and colonisation. The development of maritime transport has been both a significant driver of human development and responder to changing sustainability challenges throughout history. These transport challenges continue today with the additional real concerns as to climate change, marine pollution and potentially significant movement of human populations.

Some important stages in the development of maritime transport may be identified and linked with the main strands of sustainable human development. These are summarised in Table 8.5.

TABLE 8.5 Stages in ship and boat development (National Maritime Museum, 1981)

SHIP/BOAT DEVELOPMENT	*LINK TO SUSTAINABLE DEVELOPMENT*	*TIME/LOCATION*
First use of log or skin boats or rafts. Hand-paddle-driven	Early spread of human populations to Australasia and the pacific.	Since 10000–30000 BCE. Europe, Asia
Multi-oared vessels with single square sails	Early trade route development. Transfer of knowledge and technologies between East and West.	From around 1000 BCE
Robust clinker-built Norse trading vessels, single square sails	Development of Northern European trade routes. European settlements in Iceland, Greenland and North America.	From around 500 to 600 CE
Specialist single-masted purpose-designed trading vessels (cogs)	Development of regularised trading associations (e.g. the Hanseatic League). Some initial exploration down African coast.	From around 1200 to 1400 CE
Development of two to three masted sailing vessels with at least one fore- and aft rigged mast allowing sailing closer into the wind	More efficient transport of freight; initiation of trade and fishing links with the Americas.	From around 1400 CE
Development of more efficient hull forms and sail plans culminating in highly efficient clipper ship designs	More efficient and more reliable freight transport. Developing countries expand economies through increased export of manufactured goods.	From around 1700 to 1900 CE
Initial development of practical coal-fired steam engines from previous crude land-based pumping engines	Initially only support to sailing vessels as tugs or very short sea crossings. Improved turn-around times in port.	1800–1820 CE
Rapid improvement of compound steam engine efficiency and the development of screw as opposed to paddle power challenges the sail supremacy	Allows increased reliability of steam vessels and regularisation of trade and migration routes. Increased pollution and GHGs from coal-fired engines.	1840 onwards

8.8.2 Emissions and Environmental Impacts

Although ships are the most energy-efficient method to move a given mass of cargo for a given distance, the sheer size of the industry means that it has a significant effect on the environment. Maritime freight is responsible for about 10% of GHG emissions, whilst air travel is by far the most polluting mode per tonne transported (1,054 gCO^2/tonne-km as compared to 19 gCO^2/tonne-km); if done carefully, the shift from air to sea can significantly reduce freight emissions without compromising deliveries.

In a Review of Maritime transport 2024, The United Nations Conference on Trade and Development points out that in this sector, GHG emissions have risen 20% over the last decade, with aging fleets of vessels that run almost exclusively on fossil fuels (UNCTAD, 2024).

Optimistically, the International Chamber of Shipping (ICS, 2022) takes the view that shipping is the least environmentally damaging form of commercial transport and, compared with land-based industry, is a comparatively minor contributor to marine pollution from human activities. It claims that there has been a substantial reduction in marine pollution over the last decade and a half, especially with regard to the amount of oil spilled into the sea, despite a massive increase in world seaborne trade.

The ICS also asserts that the shipping industry is a small contributor to the total volume of atmospheric emissions compared to road vehicles and air transport due to significant improvements in engine efficiency and hull design over the last decade. The use of ships with larger cargo carrying capacities has led to a reduction in emissions and an increase in fuel efficiency.

In support of this, in terms of CO_2 emissions per tonne of cargo transported one mile, shipping is recognised as the most efficient form of commercial transport, although the quality of the fuel used often causes serious local pollution problems. However, the enormous scale of the industry (80% of world goods trade) means that it is nevertheless a significant contributor to the world's total GHG emissions with around 3% of total global GHG emissions (UNCTAD, 2024).

8.8.3 Way Forward

The decarbonisation of shipping as a contribution to lowering global GHG emissions is an ongoing goal to reduce GHG emissions from shipping to net zero by or around 2050, championed by the International Maritime Organization (IMO). The IMO has an initial strategy that includes lowering, or limiting, the combustion of fossil fuels for power and propulsion to limit emission of carbon dioxide (CO_2); development of alternative fuels such as green ammonia, hydrogen and biofuels to reduce reliance on fossil fuels; adopting digital technologies to increase vessel efficiency; and reducing emissions by 40% by 2030.

However, maritime transport is also one of the economic sectors with a greater potential to reduce its GHG emissions. Establishing a roadmap to the decarbonisation of the maritime sector will require a consideration of three different fundamental aspects:

technology, policy and investment in research, development and innovation. Options for reducing maritime transport carbon emissions include the following:

- Low-carbon fuels
- "Slow steaming" (a strategy aimed at reducing fuel consumption and emissions by intentionally reducing engine power)
- Replacement of aging vessels with newer more energy-efficient designs
- Returning to advanced-technology wind power

Decarbonisation fuel options for shipping still requires research and development, though advanced biofuels, ammonia and synthetic fuels are emerging as viable options. In the meantime, interim measures are being researched. For example, there is a recently reported proposal to use limestone pellets to scrub most of the carbon from a ship's exhaust. The CO_2 in the exhaust reacts with the calcium carbonate (limestone), forming bicarbonate minerals that can remain stable in the ocean (Bryce, 2025).

8.9 INTRODUCTION TO PORT DEVELOPMENT

8.9.1 Background

From a civil engineering standpoint, the size of vessels dictates the capacity of ports, which means that smaller harbours are restrictive and that countries which depend on maritime trade also have large ports to serve main container traffic. The ports engineer has a big range of facilities to look after nowadays and must consider the impact on the environment and the sustainability of the overall operation.

Port sustainability encompasses the intention of developing and operating without having an impact on your surroundings in any form. At present, this is taken to mean not depleting what are identified as finite resources and not having a human impact on the natural process of climate change.

Proving a negative is impossible. It can never be proven that any development or operation is sustainable. There can only be a degree of compliance with the intention, and that degree of compliance depends on the perspective of the decision maker. What priorities and proofs do the decision maker wish to apply to their definition of sustainability?

Setting aside the philosophy of sustainability, it is more practical to identify, prioritise and manage the development and operation of a port so they are consistent as far as practical with their environment and minimise the change they contribute to that environment. To assist with this the impact of ports on their environment can be categorised into five areas:

1. Physical changes the development of a port has on the geography, oceanography, meteorology and biology of its surroundings.
2. Interruptions and changes within the water cycle caused by the operation of the port. Where and how is clean water sourced, and how is effluent water managed, treated and put back into the port's surroundings.
3. Interruptions and changes to the atmosphere primarily from the production and consumption of energy for port operations (emissions both particulate and gaseous) and also covering noise and other changes.
4. Effects due to the operations of port users including ships, road haulage, rail and other aspects of the supply chain. These are interlinked with the previous three categories.
5. Accidents, disasters and associated response. These are interlinked with the other four categories.

This section investigates best practices in each of these areas outlining how they can be addressed and the standards that can be achieved.

As a check to the relevance of these categories, they can be benchmarked against issues identified by the World Association for Waterborne Transport Infrastructure (PIANC) indicated in its report "Sustainable Ports – A Guide for Port Authorities" (PIANC, 2024), which were as follows.

1. Environmental quality (soil, water, air and noise)
2. Habitat and integrity of ecosystems
3. Energy efficiency and energy transition (from fossil towards clean fossil towards renewables)
4. Materials and waste management
5. Climate change mitigation and adaptation
6. Stakeholder participation and corporate social responsibility
7. Co-operation with private sector, public authorities, NGOs, academic world and other ports.

The parallels are clear for categories 1–3. Category 4 is the practical outworking of stakeholder participation and the broader community involvement. Category 5 has perhaps not been addressed by the PIANC report, though it is implicit in climate mitigation, adaption and broader community responsibilities.

The approach a port takes to the management of its impact can be considered in the following three stages:

1. The impacts the port has direct control over (captive power, tugs, etc.)
2. Impacts where the port can set standards and manage the activity of others to adjust them (tenants, suppliers and operators)
3. Secondary impacts where the port can set an example but has no significant control and limited influence (mains power generation, external potable water, rail connections, etc.)

It is helpful for port management to understand and monitor all impacts through this approach and thus ensure its efforts are focused on the impacts it can effectively manage.

8.9.2 Physical Changes Caused by the Development of a Port

A port is the interface between land and sea. It is where people and cargo get on and off ships and boats so they can travel between islands and continents. To do this, it is essential that a quay or jetty is built where a ship or boat can float adjacent for ease of access. In building facilities to do this, it is inevitable that there will be an impact of coastal processes including erosion and deposition caused by wind, waves and currents, and drainage of water from land to sea.

The impact of these and the change of use of land, intertidal area and near-shore seabed has an impact on the ecology of the coastal area.

The impact of the development of the port depends on how large the development proposed is, the sensitivity of the section of coast where it is developed and how well the development is planned to mitigate the impacts that it will cause, in that it must be remembered that decisions will be made on which impacts are more acceptable.

A further element complicating port developments is the long-term nature of such developments. The term "the 100-year port" refers to an often quoted need to plan ports for the next 100 years. Such a long-time horizon makes a fool out of all forecasts in 1923 when the main bunker fuel for shipping was coal.

Best practices for planning and design are as follows:

1. Plan for 100 years but build for the foreseeable future based on contracts and the level of commercial assurance of volume and revenue
2. Reuse, repurpose and adapt existing facilities where this is practical
3. Prepare a detailed forecast of the performance requirement for the port, which covers how many ships and boats will call, their size and in the context of the location, the chandlery, repair and bunker needs; the volume and nature of that volume (what it is, how it is carried, special characteristics); and an understanding of the place of the port in the supply chain and the consequent impact on storage and hinterland connectivity.
4. Fit the port into the broader planning for the region ensuring that key steps in the logistics process are encouraged to take place at the optimal location
5. Optimise productivity in terms of gross handling productivity, security/safety of operations, multipurpose functionality, etc.

8.9.3 Ports and the Water Cycle

The interaction of ports with the water cycle is driven by three main factors:

- Demand for potable and high-purity water
- Process and wash water
- Contamination of surface water

The simple and best practice ideal solution is all water entering the site, potable, groundwater, surface water, precipitation, etc. must leave the site no more contaminated than when it enters the site. This requires the following:

1. All water entering the site must be identified and monitored.
2. Where there is the potential for port development or operations to impact the water entering the site, methods of intercepting it must be put in place.
3. Depending on the potential impacts, the water may need storing, treating and testing before it is allowed to leave the site. The order and balance of the testing, storage and treatment will vary depending on the potential impact.
4. Adequate and suitable sanitation arrangements should be done.

The practical implementation of this simple, best practice, solution has some qualifications that should be understood. Water in nature is rarely pure, water (H_2O), in the atmosphere, on the ground surface and through the subsurface it collects minerals, dissolved gases, particulates and other additives. These can be perceived as positive contributors (the value of "mineral water") or negative contributors (pollutants). Setting a standard for the release of water from a port into its environment should involve consideration of the original needs of the environment of nutrients, minerals and other additives.

Being more specific, the following are practical measures that should be considered during the planning, development and operation of any port:

1. Monitoring for all water courses (river and stream), groundwater, etc. relevant to the ports site.
2. Impermeable barriers between structures, operating areas, processing areas and all water courses and groundwater; this includes open storage; however, with open storage, further consideration should be given to the nature of drainage and collection.
3. Collection and treatment or storage and treatment of all water before it is released back into the port's environment.
4. Careful selection and monitoring of the location and method of release for treated water into the port's environment.

8.9.4 Emissions-Related to Ports

The glow in the sky from an active container terminal operating 24/7 and the clanging of containers as they land and spreaders release their twist locks is perhaps one of the most enduring memories of those living close to a container terminal. It illustrates that whilst gaseous and particulate emissions are critical matters, emission impacts from ports to their environment are far more varied in their nature. Owls and other nocturnal birds and animals are as confused as whales and dolphins by noise and light in their environment.

The primary sources of gaseous and particulate emissions in ports are as follows:

- Energy generation (and consumption)
- Cargo operations (container, grain, coal, etc.)
- Ships, boats and other users (addressed in other sections)
- In port industry (refineries, cold storage, final delivery preparation)
- Port-located industries (refineries, smelters, process industries, etc.)

Sustainable energy generation and use in a port is a subset of the broader discussion of sustainable energy across all human endeavour. There are, however, some specific features related to the uses of power in ports that add complexity to the delivery of that energy in a sustainable manner. In essence, these features are as follows:

- High electrical loads related to high accelerations for heavy lifts (20–40 t) during cargo handling operations
- Variable levels of base load demand for refrigerated storage, lighting and other operations

In addition to cargo lifting operations, the horizontal transport of cargo within a port requires energy. This has primarily been undertaken by road haulage vehicles. Within a port, the use of electric versions of such vehicles is more practical than where those same vehicles need to travel long distances away from their base of operations. The range, recovery and recharging of EVs have less impact on the conversion of port operations than the conversion of most other uses of road haulage.

At times ports can provide excellent locations for the generation of sustainable power for captive use or to feed other users. Examples of this include breakwaters as sites for wind turbines and current turbines.

Road haulage is not the only option for horizontal transport within a port. Conveyors and pipelines are commonly used and commonly can be powered by sustainably generated electricity. Less commonly, rail systems can be used.

Light and noise emissions also need to be addressed within the port environment. Automation can contribute to the reduction of light emissions. However, this is best addressed through the following:

1. Careful placement and modelling of lighting schemes to ensure they minimise the external influence of the port's lights.
2. Effective close-out practices to shut down lights when they are not required. As an example, security does not require visible light and other active systems. Passive heat sensors, ultraviolet and sound systems can be as effective and less aggressive and provide excellent covert information to aid the maintenance of security.

It should be remembered that light emissions seaward can be as big an impact as that toward landward neighbours.

Noise is a surprisingly complex emission for a port to handle. Noise can propagate through the atmosphere and water. The steps to reducing impact are as follows:

1. Addressing noise propagation in preparation and adjustment of operational layout of port
2. Considering noise propagation during the assignment of berths
3. Marine and road traffic management
4. Noise control on equipment
5. The introduction of noise barriers (to be balanced with visual amenity)

8.9.5 Users, Their Impacts and the Influence of Ports on Those Impacts

"Port Governance" is a complex subject, and it influences the impact a port has on the environment. That influence can most clearly be seen in who does what and therefore can control and influence the impacts the port and its users have on the environment.

Table 8.6 provides a breakdown of the governance of a port and its users in respect of development and operations. The red line indicates the most common practices. The definition of a "best" answer is impracticable because it depends on culture, aims, objectives and time frame.

Table 8.6 demonstrates that within the development and operation process, government usually has a dominant influence. This can be exercised to limit and mitigate the impact a port has during development and operation. However, it also demonstrates that the government influence over users is considerably more limited.

The main levers of influence a port authority can have on user behaviour are as follows:

- Regulatory (for safety reasons, environmental protection, etc.)
- Financial (charges for operational and support services)
- Provision and enabling of services and support to users

In the regulatory sphere, national and international agreements govern and constrain most controls a port authority can have on its users. Therefore, active participation in the forums that set the agreements and regulations is important.

The setting of charges is a key best practice method for a port to influence its users. Examples include the Port of Rotterdam that offers discounts to sustainable vessels. Further, the early provision of services that may not initially be financially viable in the private sector but that enable users to transition to more sustainable operations is an influence ports can have on their users. Port services that will enable users to transition to more sustainable operations include the following:

1. Enabling the storage and delivering to users of sustainable alternative fuels supporting their transition away from petroleum-based fuels

TABLE 8.6 Examples of port governance options indicating common practice "zone"

OWNERSHIP	*LAND*	*BASIC INFRASTRUCTURE*	*SUPERSTRUCTURE*	*QUAY OPS.*	*MARINE OPS. (TUGS)*	*LAND OPS. (CARGO)*	*USERS*
100% public ownership	State ownership of port land is common practice though that ownership can be vested in different levels of government	Examples: • Public works dept. • Government • Port authority • State-owned companies (national, state or local)					It is rare for many users to be in public ownership. The exceptions are usually ferry terminals
Some form of division between public and private	Rare, UK and USA have examples	Joint ventures between port authorities, state-owned companies and Private companies Concessions or licenses from Government or Port Authorities to private companies Private stevedores using common berths					
100% privately owned	Rare, UK and USA have examples	Rare		Within common practice		Common practice	Dominant practice

2. Provision of shore power (cold ironing)
3. Reception of waste materials (wastewater(s)), chemicals (batteries) and solid waste
4. Inspection, maintenance and repair facilities including hull cleaning to improve ships' hourly fuel consumption whilst at sea

8.9.6 Disaster and Emergency Response

Accidents *should not* happen, but they do, and focussing on their prevention or avoidance is far more preferable than preparation for response. Natural disasters cannot be prevented but planning can mitigate their impact and appropriate response can add to the resilience of the community.

When accidents or disasters occur, in best practice, port authorities have a critical role to play in responding. This is consistent with a port's role as a hub or node in the supply chain. The ability to respond effectively, the resilience of a port, is an element of sustainability for both port and host community. As an example, if a fishing port can continue to operate or return to operation quickly, it delivers focus, purpose and income to impacted communities.

In line with best practices, a sustainable port must be able to respond to a wide range of accidents, disasters and unexpected events. The primary responsibilities of the port are as follows:

1. Ensure it can continue to function (or return to functioning) as effectively as practical
2. Identify and limit the impact of any accident or disaster
3. Clean up, repair and restore any damage (or impact) that has occurred because of an accident or disaster

In this process, the concept of "build back better" needs to be considered. The interpretation of this concept in practice may be complex and should not be allowed to interfere with the initial restoration of the utility of the port for users and any responders to any broader aspects of the disaster or accident.

The ability for a port to respond is very significantly enhanced by planning, training and exercises. This may need to be supported where appropriate by the acquisition of backup and response equipment. To prepare, the port needs to have identified the primary challenges it may need to face during an accident or disaster. There are two ways of approaching this:

1. Identify the potential for accidents and disasters and preparing for those events
2. Understanding the assets under management and how, if they are damaged, destroyed or their operation is otherwise impaired, they can be repaired, replaced, worked around, etc.

At times, approach (2) can be viewed as identifying redundancies in port operations and could lead to cost savings for normal operations. However, resilience is sometimes ensured by redundancy, and it should be recognised accordingly as best practice. It is not possible to schedule all the types of accidents and disasters that a port may need to address. They can be natural or manmade, limited or catastrophic, and quick or slow in evolution.

A full review of the possible range of disasters is part of best practice planning in a sustainable port. As an indication of the range of potential disasters that should be considered, potential natural disasters that ports have previously suffered include the following:

- Earthquakes
- Tsunamis
- Volcanic eruption
- Flooding due to tidal or storm surge
- Flooding due to surface water runoff
- Epidemics (e.g. Ebola, COVID)

Man-made accidents and disasters a port should consider include the following:

- Marine incidents including collision, capsize, fire on board
- Catastrophic explosions of cargo on board ship
- Spills into water courses and other contamination events (from the port and its users or others)
- Toxic vapour releases into the atmosphere (for the port, its users and from external sources)
- Onshore fires that can lead to explosions
- Onshore cargo explosions
- Strikes (related to the port and external organisations)
- Riots, terrorism, insurrection and war

8.9.7 Managing Climate Risk for Ports

The climate risk to ports is a separate, although related, issue to sustainability. They may be related but a changing climate will have an impact on a port no matter how sustainable that port may consider itself to be or how sustainable others perceive it to be.

Ports are, to meet their function in low-lying predominantly coastal locations, situated in sites that have significant vulnerability to climate change. There is a drive to situate them in less-exposed coastal locations to ensure that cargo operations are optimised and ships safety is not compromised even in severe weather. However, as vast majority of ports are still typically located along open coastlines or in low-lying estuaries and deltas.

Extreme weather events are therefore a primary concern; they can damage port infrastructure and stored cargo, by flooding container storage, electrical substations, refrigerated storage, etc.; failing, vulnerable ship-to-shore cranes; and stripping roofs and insulation from covered storage.

Most if not all this damage should be eliminated by the selection of the appropriate design parameters and safety factors. With climate change, it is likely that the severity of extreme weather may increase, making current design parameters inadequate. It is also possible that the available data record is inadequate to predict the range of extreme events that ports should be designed to withstand. Whatever the reason if design parameters need to be adapted, such decisions need to be implemented as soon as practicable. There is also a legacy issue, existing infrastructure that has been designed to, now, inadequate design parameters. This may need to be upgraded to survive more extreme design parameters (if this is practical).

Extreme weather is not the only concern; its impacts can be exacerbated by any change in sea level. If mean water level increases, it compounds the situation with the impact of extreme weather. It makes it more difficult, for example, for surface water to drain off the land. The safe design level for a port and its storage therefore becomes critical, with decisions on this issue impacting on port operating cost and productivity. The larger the difference between mean water level and the operating level of the port, the more difficult (and costly) the cargo operation of the port. Existing ports when they review their actual levels have several factors to consider, including the potential that the port has subsided or been subject to geological changes in level. Legacy issues related to operating level of a port usually pose very substantial problems when the need and cost of remedial works are estimated and understood.

It should be remembered that ports are, through their exposure to flooding from surface water runoff, vulnerable to changes within the catchment areas of rivers and other watercourses that pass through or close to them. If there is significant upstream deforestation or urbanisation in the catchments of rivers, the speed of runoff increases and the risk of flooding due to surface water increases.

8.10 ADDITIONAL TRANSPORT MODES

8.10.1 Relevance

The principal focus of this chapter has been using road and marine freight transport to exemplify important transport sustainability issues. It is, however, impossible to ignore the importance rail and aviation in a discussion on transport sustainability, and this section briefly highlight this by summarising some important issues. This is done principally by summarising their historical and future development in the context of their emissions. As noted previously, around 23% of GHG emissions are contributed by transport of which 70% of direct transport emissions came from road vehicles, while 1%, 11% and 12% came from rail, shipping and aviation, respectively (IPCC, 2023).

8.10.2 Rail Transport

Rail transport has been around for almost two centuries and arguably could be said to be the first driver of energy-powered transport in the Industrial Age in Europe. Rail transport has evolved remarkably over since early 1800s. An historic review of rail transport development (Tostes et al., 2024) considers this under three time periods: the steam age (1840–1930), a transition age and the diesel electric age (1930–1920) (Table 8.7).

Rail transport is now a major worldwide means of moving freight (including bulk goods) and passenger traffic, and is a major contributor to sustainable development. Much of the world is well covered by hundreds of thousands of miles of rail track, a far cry from the first track from Stockton to Darlington in the north of England with the train travelling at a maximum speed of 10 mph. Nowadays, specialist passenger trains can travel comfortably at over 200 km/h and challenge the use of short-haul air traffic, both in terms of convenience and emissions.

Significant advances have also been made with heavy freight trains, particularly in the transport of mined mineral ores or coal. Table 8.8 indicates the importance of rail goods traffic in range of countries with contrasting transport environments. In some countries, the direct mine to marine port transport option is perceived as a highly

TABLE 8.7 Key periods in rail transport development

PERIOD	*KEY CHARACTERISTICS*	*DEVELOPMENT ISSUES*
1840–1930	Initially large amounts of coal were burned in very inefficient steam engines, but then a period of great change with increasing speed, reliability and innovation. Some countries (USA) predominantly use wood until around 1900.	Rapid growth in freight and passenger traffic alongside rapidly increasing track length and coal consumption and consequent emissions.
1930–1970	Increasing cost of coal encourages rail sector to invest in diesel-electric trains. Through this period, these two energy sources become dominant. Increasingly cars, trucks and buses compete with trains, in many countries, such as the USA and UK.	Freight and passenger service continued to grow steadily on the existing infrastructure with consequent emissions. The energy efficiency of diesel locomotives was estimated as being almost 5 times greater than steam locomotives, while for electric trains, the efficiency was 15 times.
1970–2022	After 1970, most countries, apart from India and China, had already significantly reduced their coal consumption for rail purposes. These two take 25 more years. Some countries (e.g. USA) heavily invested in diesel-electric trains, while others (Japan) focused on electrification.	The CO_2 and other emissions of rail transport have reached their lowest levels in history aided by adoption of renewable energy for electricity generation.

TABLE 8.8 Freight transport changes in the last 20 years (World Bank, 2023)

COUNTRY	*2005*	*2021/2022*	*% INCREASE/DECREASE*
	FREIGHT GOODS TRANSPORTED (MILLION TON-KM)		
Australia	184,000	453,000	+146
China	2,072,600	3,018,000	+46
France	40,000	36,000	***−10***
India	407,398	719,620	+77
Indonesia	7,000	15,700	+124
Poland	50,000	55,000	+10
Russian Fed.	1,852,913	2,638,562	+42
UK	21,426	15,212	***−29***
USA	2,476,653	2,239,400	***−10***

cost-efficient option, with use of trains frequently over 100–200 wagons not unusual. These massive trains, powered by multiple locomotives, are designed to efficiently transport large volumes of ore across vast distances. In fact, a BHP ore train used to transport iron ore in Western Australia, is credited with the record of 682 wagons and a length of 7.3 km.

Looked at in isolation, the increase in electrification on rail networks is encouraging from an emissions perspective. Within the sector, the carbon intensity of useful energy in electric trains is 16% lower than that of diesel trains. However, when considering transport overall, the picture is not so good, as the overall share of rail continues to decline in comparison to CO_2-intensive transport modes such as cars, trucks and planes.

8.10.3 Aviation Transport

The aviation sector has grown impressively, particularly in the past 50 years, and this may be summarised in number of periods (Table 8.9). The air transport sector is primarily concerned with passenger traffic, as the freight element is less than 1% that of the rail sector in terms of tons-km. Although some decline in numbers of people flying have been witnessed in the past few years due to COVID, the numbers have now rebounded (Figure 8.10).

GHG emissions are a growing concern with regard to air transport, which is not altogether surprising since the fuels used in flying are virtually all fossil-based. Although some experimentation is taking place with biofuels, it will be some time before the aviation sector can switch to cleaner fuels and reduce significantly the reliance on fossil fuels.

Although planes have had the capacity for six decades to travel at speeds greater than Mach 1.0 (the sound barrier), people have learned to accept these speeds are just not commercially viable or environmentally acceptable. Lower speeds of around 600 km/h and longer flight times are accepted. Although a British/French design of the 1960s, the Concorde, flew commercially, successfully for 27 years at supersonic speed (over 1,300 mph) cutting the journey times in half across the Atlantic between New York and

TABLE 8.9 Historic development of passenger aircraft

PERIOD	*CHARACTERISTICS*	*DEVELOPMENT ISSUES*
1903–1914 Pioneer period	Principles of powered flight established. Aviation passed from the preserve of eccentric enthusiasts to being an established technology, with professional aeronautical engineering organisations. Different aircraft configurations were experimented culminating in the basic design of a petrol-engine tractor biplane emerging as a common option. Airship development.	Gradual spread of technology from USA to Europe and beyond. No established passenger routes. Initial experiments with mail carrying. Airships use flammable lighter than air hydrogen gas.
1914–1918 WWI	Establishment of military plane usage. Increase in plane speeds and size. Multi-engine aircraft introduced.	Development of specialisations such as reconnaissance, fighter and bomber aircraft. Airships used in air-bombing and civilian casualties become a norm.
1918–1939 Inter-war development	Inter-war period saw application of war technology in civil aircraft. Evolved from wood and fabric low-powered biplanes made to sleek, high-powered aluminium monoplanes in the first true airliner development.	Longer range flights begin to establish realistic and reliable passenger travel including commencement of intercontinental travel initially by rising large flying boats. Hazards of airships for passenger traffic exposed by high-profile disasters (R101 and Graf Zepplin).
1939–1945 WWII	World war again gives great impetus to major advances in aircraft technology in terms of size, speed and endurance. First practical application of the jet engine.	Population as a whole put at risk by development of heavy long-range bombing capabilities with minimal efforts paid to keeping to military targets.
1945–1970 Post-war development	Post-war application of war technology in civil aircraft. Introduction of piston engine airliners followed shortly (1950s) by jet and jet-turbine-powered craft. Aircraft such as the redesigned Comet and the Boeing 707 spearhead the use jets in commercial aviation.	Aviation expands quickly, relying initially on former military aircraft to carry passengers and cargo. The DC-3, for example, played a long and key role, enabling efficient air transport worldwide. Expansion of airlines make flying an increasingly affordable and popular travel option (Figure 8.10). High-profile crashes initiate a greater focus on safety.
1970–2000 Popular expansion	Increased range and size of passenger aircraft – focus on size and speed in the "jumbo/super jumbo" concept and on the supersonic Concorde.	Increasing numbers of aircraft, airlines and passengers as flying becomes increasingly popular for leisure travel. Air travel rapidly overtakes marine passenger shipping which largely disappears apart from short-range ferries and cruise ships. Increasing concerns as to noise and emission pollution.

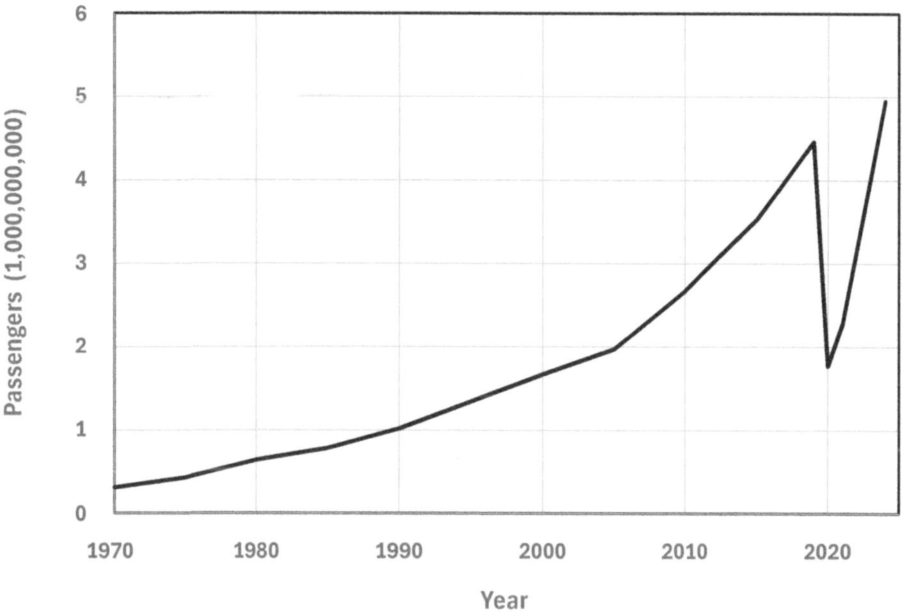

FIGURE 8.10 Growth in air passenger traffic (World Bank, 2024).

London/Paris. Eventually, the costs and environmental constraints were just too much, and the Concorde operations ceased in 2003.

Sometime in the future, assuming that technological advances allow, there ought to be supersonic flight available as standard, possibly before the end of the century. However, the main thing that will be looked at to solve in the immediate term will be the GHG emissions and noise impacts. If the biofuel that is found is readily available and GHG can be reduced without penalising faster travel, then that could be the next stage in aviation development.

8.11 SUMMARY

Transport has been at centre of our development from the earliest ox-drawn sleds and wooden rafts to the latest high-speed trains and reusable space shuttles. The increasing price to pay for developments in the last two centuries has been in the damage to our environments on land, sea and atmosphere. In the last three decades, there has, in particular, been a focus on GHG emissions and the impact on short- and long-term climate.

As noted in the above sections, transport is a high-profile target for criticism with respect to GHG emissions. In the UK, transport is the largest culprit in terms of GHG emissions, producing around 28% of the country's total emissions. Globally, the figure is similar with 95% of the world's transport energy still coming from fossil fuels.

In the future, transport will have to be less focussed on speed, size and national pride and more on efficiency and the use of clean or renewable energy. However, the Sixth IPCC Report (2023) is somewhat pessimistic. It warns that meeting climate mitigation goals would require transformative changes in the transport sector. However, emissions from shipping and aviation continue to grow rapidly and that as a whole, transport-related emissions in developing regions have increased more rapidly than in Europe or North America, a trend that is likely to continue in coming decades.

In response to public pressure, it must be the responsibility of national, regional and local governments to produce transport plans with a climate change focus based around low-carbon transport sustainable pathways that contain realistic and agreed targets for decreasing GHG emissions.

REFERENCES

ADB (2003) *Environment Assessment Guidelines.* Asian Development Bank, Manila. https://www.adb.org/sites/default/files/institutional-document/32635/files/environmental-assessment-guidelines.pdf

ADB (2014) *Climate Proofing ADB investment in the Transport Sector: Initial Experience.* Manila.https://www.adb.org/publications/climate-proofing-adbinvestment-transport-sector-initial-experience

ADB (2020) *Manual for Climate Change Adjustments for Detailed Engineering Design of Roads. Experience from Viet Nam.* Wilby, R. for Asian Development Bank. https://dx.doi.org/10.22617/TIM200147-2

Alcaraz, M. and Rigaud, K. K. (2023) *Embedding Climate Resilience into Energy Sector*

Austroads (2017) *Guide to Pavement Technology Part 2: Pavement Structural Design.* AGPT 02-17. Austroads, Sydney. Available at: https://austroads.gov.au/publications/pavement/agpt02-17

Austroads (2020) *Sustainable Roads Through Fit-for-Purpose Use of Available Materials: Evaluation Tool and User Guide.* Austroads Ltd., Sydney.

BBC (2024) *Wireless Charging: The Roads Where Electric Vehicles Never Need to Plug In.* Available at: https://www.bbc.com/future/article/20240130-wireless-charging-the-roads-where-electric-vehicles-never-need-to-plug-in

Bryce, E. (2025) *Shipping is One of the World's Dirtiest Industries: Could this Invention Finally Clean Up Cargo Fleets?* Guardian. Available at: https://www.theguardian.com/environment/2025/jun/26/global-shipping-emissions-invention-clean-up-cargo-fleets-net-zero

Cook, J. R. (2022) Reducing future climate impact on rural transport infrastructure in developing countries; the role of engineering geology. Quarterly Journal of Engineering Geology and Hydrogeology. Available at: https://qjegh2021–052?ref=pdf&rel=cite-as&jav=VoR

Cook, J. R. and Petts, R. C. (2023a) Road classification and function. Chapter 3 in *Rural Road Engineering in Developing Countries.* CRC Press, pp. 25–36.

Cook, J. R. and Petts, R. C. (2023b) Road development and related policy. Chapter 1 in *Rural Road Engineering in Developing Countries.* CRC Press, pp. 1–12.

Cook, J. R. and Petts, R. C. (2023c) The rural road environment. In *Rural Road Engineering in Developing Countries.* CRC Press, pp. 75–88.

Cook, J. R. and Petts, R. C. (2023d) Road maintenance strategy. In *Rural Road Engineering in Developing Countries.* CRC Press, pp. 355–376.

Cook, J. R. and Petts, R. C. (2023e) Appraisal of low volume rural road initiatives. In *Rural Road Engineering in Developing Countries*. CRC Press, pp. 61–74.

Cook, J. R., Petts, R. C. and Rolt, J. (2013) *Low Volume Rural Road Surfacing and Pavements: A guide to Good Practice*. Research Report for AfCAP and UKAID-DFID. Available at: https://www.research4cap.org/ral/Cook-etal-Global-2013-LVPGuideline-AFCAP-v130625.pdf

Cook, J. R., Petts, R. C., Visser, C. and You, A. (2017) *The Contribution of Rural Transport to Achieve the Sustainable Development Goals*. Research Community for Access Partnership (ReCAP) Paper, ref. KMN2089A, for UKAID-DFID. Available at: https://www.research4cap.org/index.php/resources/rural-access-library

Davies, E. H. (2011) *Roads in Roman Britain*. The History Press. Cheltenham, UK.

Gillespie, W. M. (1850) *A Manual on the Principles and Practice of Road Making*. A. S. Barnes & Co., New York, Available at: https://archive.org/details/manualofprincipl00gillrich/page/36/mode/2up?view=theater

ICS (2022) *Environmental performance: Comparison of CO_2 emissions by different modes of transport*. International Chamber of Shipping. Available at: https://www.ics-shipping.org/shipping-fact/environmental-performance-environmental-performance/

IPCC (2023) Sixth assessment report. In *Transport*. Available at: https://www.ipcc.ch/report/ar6/wg3/chapter/chapter-10/

Lay, M. G. (1992) *Ways of the World: A History of the World's Roads and of the Vehicles That Used Them*. Primavera Press, Sydney.

National Maritime Museum (1981) *The Ship* (Volumes 1–10). HMSO Stationary Office, Wisbech, Cambridgeshire, UK.

Neal, I. (2012) The potential of sand dam road crossings. *Dams and Reservoirs* 22(3–4):129–143.

Parkman, C. (2014) *Rapid Assessment of Research Gaps in Road Engineering and Technical Aspects*. High Volume Transport Project Report for DFID. Available at: https://www.gov.uk/research-for-development-outputs/high-volume-transport-rapid-assessment-of-research-gaps-in-road-engineering-and-technical-aspects

PIANC (2014) *Sustainable Ports: A Guide for Port Operators*. World Association for Waterborne Transport Infrastructure. Available at: https://www.pianc.org/publication/sustainable-ports-a-guide-for-port-authorities/

Pryor, F. (2004) *Britain BC*. Harpercollins, London, UK,

ReCAP (2019) *Economic Growth through Effective Road Asset Management (GEM)*, Final Report, Project No. 10636A GEN2018A. London: ReCAP for DFID. Available at: https://www.research4cap.org/index.php/resources/rural-access-library

Salman, M., Bunclark, L., AbuKhalaf, M., Borgia, C., Guarnieri, L., Hoffmann, O., Sambalino, F., van Steenbergen, V. and Lebdi, F. (2016) *Strengthening Agricultural Water Efficiency and Productivity on the African and Global Level: Status, Performance and Scope Assessment of Water Harvesting in Uganda, Burkina Faso and Morocco*. FAO, Rome, p. 59. Available at: https://openknowledge.fao.org/server/api/core/bitstreams/16ceb4a2-f72a-4e7b-aea2-c70528e448b9/content

Sasidharan, M., Torbaghan, M. E. and Burrow, M. P. N. (2019) *Using Waste Plastics in Road Construction*, K4D Helpdesk Report. Institute of Development Studies, Brighton, UK. Available at: https://opendocs.ids.ac.uk/opendocs/bitstream/handle/20.500.12413/14596/595_Use_of_Waste_Plastics_in_Road_Construction.pdf?sequence=62&isAllowed=y

Simpson, J. and Clark, A. (2023) *Pothole Damage from Electric Cars is Double that of Petrol*. Daily Telegraph. Available at: https://www.telegraph.co.uk/news/2023/06/26/pothole-electric-cars-damage-roads-double-petrol-telegraph

Skempton, A. W., Chrimes, M. M., Cox, R. C., Cross-Rudkin, P. S. M, Rennison, R. W. and Ruddock, E. C. (2015) *A Biographical Dictionary of Civil Engineers in Great Britain and Ireland: McAdam*. ICE and Thomas Telford Limited, London.

SuM4All (2024) *Building Momentum for Global Impact. The Role of Transport in Sustainable Development.* Sustainable Mobility for All. Available at: https://www.sum4all.org/news/building-momentum-global-impact-role-transport-sustainable-development

Tostes, B. A., et al. (2024) On the right track? Energy use, carbon emissions, and intensities of world rail transportation. *Applied Energy* 367:123344.

Towler, J. and Dawson, J. (2009) History of chip sealing in New Zealand. In *1st ARRB Chip Seal Conference*, 28–29 July 2008. Adelaide.

TRL (2023) *A Guide to Structural Design of Surfaced Roads in Tropical and Sub-tropical Countries*, RN 31 (5th Edition). TRL Ltd for DFID, UK.

United Nations (2016) *Mobilizing Sustainable Transport for Development.* Analysis and policy recommendations from the United Nations Secretary-General's High-Level Advisory Group on Sustainable Transport. Available at: https://sustainabledevelopment.un.org/content/documents/2375Mobilizing%20Sustainable%20Transport.pdf

UNCTAD, (2024) *Review of maritime transport, 2024. Navigating maritime chokepoints.* United Nations Conference on Trade and Development. United Nations, Geneva. Available at: https://unctad.org/system/files/official-document/rmt2024overview_en.pdf

Van Steenbergen, F., Woldearegay, K., Agujetas Perez, M., Manjur, K. and Al-Abyadh, M. A. (2018) Roads: Instruments for rainwater harvesting food security and climate resilience in arid and semi-arid areas. In Leal Filho, W. and de Trincheria Gomez, J. (eds.), *Rainwater-Smart Agriculture in Arid and Semi-Arid Areas.* Springer International Publishing. Available at: https://roadsforwater.org/wp-content/uploads/2019/01/Frank-et-al-_Roads_-Instruments-for-Rainwater_-2018.pdf

World Bank (2019) *Global Roadmap of Action toward Sustainable Mobility.* Sustainable Mobility for All (SuM4All). Available at: https://documents.worldbank.org/en/publication/documents-reports/documentdetail/948891576846871505/global-roadmap-of-action-toward-sustainable-mobility

World Bank (2021) Green roads for water: Guidelines for road infrastructure in support of water management and climate resilience. Van Steenbergen, F., Arroyo-Arroyo, F., Rao, K., Hulluka, T. A., Deligianni, A. and Woldearegay, K. (eds.), International Development in Focus, World Bank, Washington, DC.

World Bank (2023) *The Impact of Infrastructure on Development Outcomes. A Qualitative Review of four Decades of Literature.* Policy Research Working Paper 10343.

World Bank (2024) *Databank. World Development Indicators.* World Bank Group, Washington, DC. Available at: https://databank.worldbank.org/source/world-development-indicators/Series/IS.RRS.GOOD.MT.K6

Civil Engineering and Sustainability

9

Scott Younger and Jasper Cook

9.1 OVERVIEW

9.1.1 Additional Sectors

Previous chapters have covered three key sectors, water, transport and energy in some detail. While these sectors are certainly crucial in terms of human development, there are other topics under the civil engineering umbrella that require some consideration in terms of their contribution to identified Sustainable Development Goals (SDGs). These are as follows:

- Human habitations/infrastructure
- Agricultural engineering
- Large structures
- Geotechnical engineering (soil–rock mechanics, engineering geology)

The provision of shelter would be uppermost, followed by the usage of the land primarily for the growing of crops, providing grassland for domestic animals or its retention as natural forest, prairie or steppe. Recently there has been a dawning realisation because of the large growth in population numbers, that some of the land has to be left wild to benefit animals or plants in order to prevent their extinction and help care for life on the planet.

DOI: 10.1201/9781003566274-9

9.1.2 Buildings

Humans like most other mammals prefer living closely together, a reason why only about 1½% of the total surface land of the planet is taken up by communities mostly dwelling and working in towns and cities and hence requiring housing and other necessary buildings and structures.

Construction and associated activities account for about 40% of energy and industrial-related CO_2 emissions globally. These activities comprise the construction and operation of buildings and the production of materials. This report estimates that operation of buildings explains about 20% of global carbon emissions, followed by the supply of materials (19%) and construction services (0.3%). About 85% of total construction emissions globally come from the use of fossil fuels in buildings and materials plants while the remaining 15% comes from process or industrial emissions related to the production of construction materials (IFC, 2023).

9.1.3 Agricultural Engineering

Civil engineering has an important role within sustainable agricultural engineering by the application of engineering principles and technologies to ensure good farming practice, whilst at the same time minimising pollution and environmental impact.

This role includes developing innovative use of machinery whilst promoting renewable energy sources and managing water and soil effectively. One example of the role civil engineering can play is in the development and use of multitask machinery for use in agriculture, road construction and transport. This approach maximises use of machinery, minimises downtime and reduces emissions resulting from manufacture of redundant machinery.

Modern wheeled agricultural tractors, for example, are a low-cost mobile power source and with various attachments can be used to substitute for heavy equipment for a proven range of tasks in the road sector, often with substantial cost savings, and flexibility of applications (Petts, 2012). This simple or intermediate equipment is typically designed for ease of maintenance and repair in the conditions typical of a limited-resource environment. Table 9.1 summarises the multitasking approach.

9.1.4 Large Structures

The civil engineer involved in structures has a wide choice of subject in which to exercise his experience – housing, multi-storey apartments or office blocks, government buildings, factories and then structures that serve the needs of people, such as water, energy and transport. These include pipelines, tunnels and aqueducts to convey water, sometimes over long distances, and dams and power stations, which require special attention against radiation if for a nuclear plant. At ports, the structures will have to take note of nearshore maritime conditions and the requirement for piling of berths, for instance. For highways, there are bridges, culverts and tunnels.

TABLE 9.1 Potential agricultural tractor applications in the rural economy (Cook & Petts, 2023)

SECTOR	*OPERATIONS*
Agriculture	Ploughing, harrowing, rotovating, haulage, land clearance and levelling, root removal, planting, seed drilling, fertiliser/pesticide, harvesting, pond/dam construction fencing (post hole boring), access roads
Forestry	Winching, loading, hauling, poling, sawing, access roads
Roads	Gravel/materials haulage, water haulage and distribution, personnel transport, fuel and plant haulage (low loader trailer or semi-trailer), towed grading, dragging, towed compaction (rubber tyred/steel roller), earthworks excavation and haulage (towed scraper), excavation, loading (front shovel), grass and bush control, spreading materials, bitumen sealing (towed bitumen/emulsion heater/sprayer), stone crushing (towed crusher and screens), recycling pavement (milling attachment), brushing/sweeping, mixing (disc harrow)
Agro-processing	Threshing, hulling, milling, haulage
Water sector (non-road)	Pipeline excavation, pipe laying, cranage, loading, earth dam construction, irrigation channel construction, water pumping, water haulage, borehole drilling
Building contracting	Materials haulage, loading
Mining/quarrying	Stone crushing (from power take-off), loading, access roads, materials haulage
Transport services	Loading, short haulage: goods, materials and personnel

The range of design methods for structures and the construction materials to be employed have changed remarkably since the onset of the Industrial Revolution. These changes can be usefully exemplified by looking at the historical approaches to bridging the rivers of the Scotland's east coast, the river Tay and the Firth of Forth (Table 9.2).

A view of the three Forth bridge structures is shown in Figure 9.1. The design of the first two bridges, far advanced for its time, would have been carried out using hand calculation based on elastic theory with some approximations based on experience. The Forth rail bridge, learning material and structural design lessons from the Tay bridge disaster, has stood the test of time, just needing paintwork against rusting applied on a periodic basis and other periodic maintenance, as required. The first Forth road bridge, a suspension structure, was constructed seven and a half decades later, when computation had advanced, the first computers were coming online, and there was a possible use of a mainframe computer for elements of the structure. The third and most recent Forth bridge, called the Queensferry Crossing, is the longest, 2.7 km, cable-stay structure in the world, for which advanced computer-based design would have been readily available. Each structure represents different periods as the profession has taken on and applied more sophistication in its design of the structures. The site of the three bridges has been awarded World Heritage status.

TABLE 9.2 East coast Scottish bridge comparison

STRUCTURE	*YEAR COMPLETED*	*COMMENT*
The old road bridge, Perth	1776	Seven-span traditional stone masonry arch bridge designed by John Smeaton using traditional methods. Still operational for vehicle traffic.
First Tay rail bridge	1878	Single-line cast iron bridge with braced columns and girders. High girder collapsed in December 1879 with loss of all on board a north-bound train during force 10/11 storm. A formal enquiry found the bridge design inadequate to withstand wind forces (contributary factors were poor construction and maintenance).
Second Tay rail bridge	1887	New double-line wrought iron columns and girders on brick and concrete foundations (3.26 km) was built approximately 60 m upstream of the previous structure. Design tested specifically for weight and wind force resilience.
Forth rail bridge	1890	Initial suspended bridge design cancelled after the Tay Bridge disaster. A new 'rigid' massive cantilever-based steel design adopted (3.26 km). First use of steel in a major transport structure (53,000 tons).
Forth road bridge	1964	A steel suspension bridge (2.52 km long) was the longest in Europe at the time. Designed to carry vehicles as well as cyclists and pedestrians. Materials: 210,000 tons concrete, 39,000 tons steel.
First Tay Road bridge	1966	A welded steel box girder and concrete deck composite dual carriageway road bridge (2.25 km). A quarter-size model was tested to destruction as part of design process. Materials: 140,00 tons concrete, 8,000 tons structural steel, 4,600 tons mild steel. Designed to carry cyclists and pedestrians as well as vehicles.
Queensferry Crossing	2017	The longest three-tower cable-stayed road bridge in the world (2.7 km). Completed under a design-and-build type of contract; the bridge carries motorcycles, cars and heavy goods vehicles, while public transport, cyclists and pedestrians use the adjacent Forth road bridge. The selected design options took into account. Intelligent Transport recommendations were adopted for adjacent access road systems to allow savings in emissions.

FIGURE 9.1 The three Forth bridges: (a) rail bridge, (b) road bridge and (c) Queensferry Crossing.

9.1.5 Geotechnical Engineering

Geotechnical engineering, or soil mechanics as it was first known, was in its early formative days as the industrialisation advanced through the 19th century (Terzaghi & Peck, 1948). In the inaugural edition of *Geotechnique* in 1948, Skempton noted that by the beginning of the 20th century, the great advances in construction had by far outstripped the old empirical design methods. He identified the way forward that was taken through the development of geotechnical engineering within existing soil mechanics and engineering geology approaches to materials and their behaviour. A key point he highlights, which is still highly valid today, is the difference between material behaviour in the field and behaviour in the laboratory. Although frequently significant, this is still not given enough attention.

The importance of well-documented and rigorously analysed case studies is still of importance today even in our new environment of virtual models and AI.

9.2 THE DIGITAL INFLUENCE

In the later 1930s, John Logie Baird produced the first black and white television while Alan Turing, working at Bletchley Park during WWII on secret codebreaking, started to develop the methods of computation which would match the first mainframe computers that appeared at universities in the early 1960s. Today, 'life' would be unthinkable without these pioneering efforts – no television, no personal computer, no putting a man on the moon, and no thinking about and applying artificial intelligence (AI).

On a personal note, Dr Younger notes he was lucky enough to have use of one of the first large commercial machines that came to some universities in the early 1960s and used it with a senior colleague to help design the outfall channel works of a large new thermal power station located on the banks of the River Forth (Smith & Younger, 1968). In those days, one had to write one's own programmes, which were basic in format (using a language from Algol, Fortran or Basic) compared with what we would be able to do nowadays. However, that experience was good enough to enable use of a variation of that software for checking, in 1979, 40 bridge foundation designs (three of them significant) for 350 km of new provincial highways in Thailand and save a significant sum on construction costs, a saving on rather conservative over-design done by hand and calculator (Younger, 1980).

But times were changing fast. Another decade saw the developments towards miniaturisation and personal computers leading on to the Digital Age. More recently, there is AI (Kirwan, 2024), which gained recognition from about 2005–2010 period. All this is within five decades, which is one reason why it is difficult to predict the overall development situation to and beyond 2100.

9.3 THE CIVIL ENGINEER

The civil engineer now and in the future is expected to be trained to understand advanced computation by computer, increasingly supported by AI. As he/she enters university it will be assumed that some knowledge of these skills will have been gained at school and even preschool. The university courses will include, for instance the principles and theories beyond elastic theory to evaluate how a foundation or structure might deform under given loading conditions and how long term wear and tear, e.g. from climate, might affect these structures. Not only will he/she learn about structures but also hydraulics, geotechnics, and many subjects such as concerning the environment, geology and economics which are likely to impinge on decision making. A few will possibly be engaged in design of structures to support space endeavours; engineers, however, will then have to contend with different gravities.

The civil engineer should be linked to his national professional body, such as the Institution of Civil Engineers in UK, which was founded in 1818, a round 30 years following the date of the start of the Industrial Age. This institution has demonstrated sustainability through its ability to embrace the various developments of society that it has had to deal with, including the after-effects of two world wars.

Historically, the positive contribution of civil engineering to human development since the onset of the Industrial Revolution has been immense in terms innovative approaches. However, in the current Anthropocene era, human activity, and particularly construction activities are consuming natural resources at an irreplaceable rate, and the question has been raised as to whether the profession of civil engineering is now doing more overall harm than good to development (Francis, 2021). To counter this suggestion, civil engineering and its associated professions perhaps need to take a more holistic approach to the long-term sustainability aspects of development, and looking at alternative to the ongoing use of concrete, steel and fossil fuels is therefore an obvious priority.

9.4 SUSTAINABLE DEVELOPMENT AND TIME LIMITS

In Chapter 1, a working definition of sustainability was established that included the phrase "with the sensible and sustainable conservation of the environment, such that future generations are not impoverished." This raises an important issue of the timescales in which we must consider sustainable development. Most famously, James Hutton as part of the Scottish Enlightenment of intellectual and scientific advancements in the mid- to late 18th century recognised the immensity of geological time, noting that he could find "no vestige of a beginning, no prospect of an end" (Hutton, 1785). We need to be aware of differing concepts of time relevant to sustainability; these are outlined in Table 9.3.

The sustainable development of an engineering structure or building is actually quite short term (Figure 9.2) in this wider concept of time.

The concept of considering sustainability well beyond traditional engineering lifetimes raises additional financial as well as immediate engineering issues. This consideration is more and more required as engineering projects become larger, the climate/environment becomes more unpredictable and the analysis for funding purposes becomes more sophisticated. Sustainability can require that an idea, project or programme, is robust enough to change with time and circumstance and maintain its substance and reflect what society requires from a profession that recognises its 'duty of care' beyond the immediate and short term.

Perhaps it is one reason the UN in devising their 17 elements of the SDGs have decided that reassessment of their aims for the Goals should take place every 15 years

TABLE 9.3 Differing concepts of time

FUTURE TIME PERIOD	*TIMESCALE*	*KEY SUSTAINABILITY ISSUE*
Immediate term	Engineering time: 100–200+ years	Lifetime target for major engineering development projects, but with potential impact into medium-term time.
Medium-term time	Short geological time: 0–1 million years	At risk of impact from engineering time emissions and pollution. Natural impacts from geological phenomena – ice ages, volcanic activity, polar reversals, etc.
Long term (deep geological time)	1 million to >1 billion years	All the vestige of current life as we know it gone. Likely complete restructuring of the earth's surface with new rock formations, resources and plate boundaries.

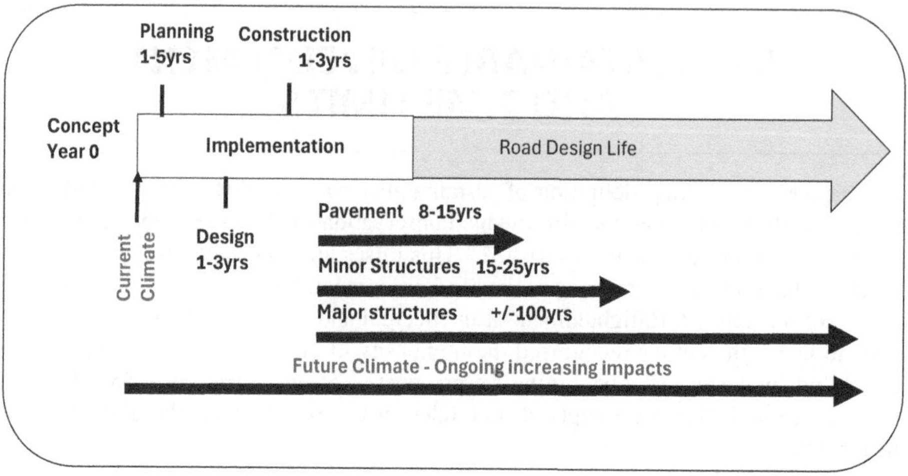

FIGURE 9.2 Engineering time in a typical road project (Cook, 2022).

and, following reassessment at the end of each 15-year period, adjust the Goals, as required. Over the four decades the Goals have been around, it is clear that many of the Goals will still require attention for several decades, perhaps centuries to come, only the substance needing alteration.

Many of them involve civil engineering inputs and, therefore, challenges where civil engineering is necessarily involved will remain for the foreseeable future. The civil engineering inputs play a necessary support function in the whole wide meaning of Sustainable Development.

In reviewing the century just passed, the predictions on where our development would be, if made a century ago, would have been quite often significantly in error. Looking back to the 1930s/1940s, the speed at which ideas have progressed from then has been truly astonishing.

REFERENCES

Cook, J. R. (2022) Reducing future climate impact on rural transport infrastructure in developing countries; the role of engineering geology. *Quarterly Journal of Engineering Geology and Hydrogeology* Vol. 55(4). Available at: https://www.lyellcollection.org/cc/climate-change-and-resilience-in-engineering-geology-and-hydrogeology

Cook, J. R. and Petts, R. C. (2023) Appropriate technology. Chapter 22. In *Rural Road Engineering in Developing Countries*. CRC Press, Abingdon, Oxford, pp. 401–422.Francis, N. (2021) Civil engineer's role in saving the world: Updating the moral basis of the profession. *Proceedings of the Institution of Civil Engineers – Civil Engineering* 174:3–9.

Hutton, J. (1785) Theory of the Earth. *Transactions of the Royal Society of Edinburgh* 1:209.

IFC (2023) *Building Green: Sustainable Construction in Emerging Markets.* International Finance Corporation, World Bank Group, Washington, DC. Available at: https://www.ifc.org/en/insights-reports/2023/building-green-in-emerging-markets

Kirwan, C. (2024) *Smart Cities; Artificial Intelligence.* Elsevier, Holland.

Petts, R. C. (2012) *Handbook of Intermediate Equipment for Road Works in Emerging Economies.* UKAID-DFID. https://www.gov.uk/research-for-development-outputs/handbook-of-intermediate-equipment-for-road-works-in-emerging-economies

Skempton, A. W. (1948) Foreward. *Geotechnique* 1(1):3–5.

Smith, A. A. and Younger, J. S. (1968) The computer aided analysis and design of a tidal channel. *Proceedings of the Institution of Civil Engineers* 10(3):305–320. Terzaghi, K. and Peck, R. B. (1948) *Soil Mechanics.* John Wiley & Sons.

Younger, J. S. (1980) Published in 6th Conference of Southeast Asian Geotechnical Society, Taipei, Taiwan.

10 Navigating Sustainability in Infrastructure Finance

Challenges and Opportunities

Euan Low and Mark Murray

10.1 WHY DO FINANCIERS NEED HELP IN NAVIGATING SUSTAINABILITY?

Mainstreaming sustainability into infrastructure capital investments requires technical experts to guide financial decision-makers to achieve optimum financial performance whilst also delivering value for all of society using our limited resources.

Our financial decision-making should prioritise capital investments in infrastructure assets and operations that avoid harm to the ecosystems essential for life on Earth. Instead, we need decision-making in capital investments to be creating value for all of society while protecting and preserving our planet's natural resources.

Too much of our capital is invested in infrastructure that is not aligned with a sustainable use of our resources. Yet, infrastructure planners, engineers, technologists and scientists have relevant subject expertise to deliver more sustainable and resilient

DOI: 10.1201/9781003566274-10

infrastructure assets and services with more robust risk management profiles that will also provide favourable financial outcomes. **How can we do better?**

10.2 WHAT CAUSES SUSTAINABILITY CHALLENGES IN THE INFRASTRUCTURE SPACE?

10.2.1 Our Resources Are Finite

We use money as a proxy for valuing our resources. Financing determines how society allocates available resources. However, current practices do not enable optimal resource allocation or sustainable resource management. Moreover, that virtual representation of money as a representation of our interactions with the real-world natural economy has become decoupled from reality and urgently needs to be corrected. Our natural assets are the ecosystems and natural resources that make all life on Earth possible. These are finite and there are biophysical limits on the 'carrying capacity' of what stresses they can tolerate.

10.2.2 What Went Wrong and How Do We Fix It?

Adam Smith's formulation of modern economic theory in *The Wealth of Nations* (Smith, 1776) was based on ideas of resource husbandry and stewardship and introduced market dynamics and the division of labour to achieve specialisation as mechanisms to increase productivity, leading to economic efficiency and growth. This led us to using money as a proxy for attributing a value on these natural assets and resources, whether these be time, materials, land, our environment, energy, intellectual property or the planet's carrying capacity. Smith's work also proposed the role of government be in protecting moral values in society, property rights, enforcing contracts and providing public goods that provided principles for modern economic theories on how environmental considerations can be integrated into sustainable finance.

Our sustainability challenges predominantly relate to our overuse of our resources through distortions of energy and material flows. Depletion of Earth's finite resources, depletion of ecosystems, increased stresses is rapidly eroding the value of our natural assets, with very evident extinction of many species and increasing socio-economic vulnerability to our own species.

To address our sustainability challenges, we need to recognise that in using money as a proxy for valuing our limited resources, then we are using monetary flows as a virtual representation of our real-world natural economy (in that traditional meaning of 'economy' being the stewardship of our natural resources). Consequently, we should be confident to assert where that virtual representation needs adjustments. And that virtual representation of money flows as a representation of our real-world natural economy has become decoupled from reality and urgently needs to be corrected. An evident example of this decoupling is in our use of fossil fuels; where the external socio-economic costs of consuming fossil fuels for energy and in material production (such as impacts from global warming, or from plastics pollution) is usually not captured in the price of fossil-fuel-based commodities.

There are difficult choices to be made in allocating these limited resources, with long-term consequences for the shape of our built environment that determines how our communities live and whether those lifestyles are sustainable within the real-world natural economy. Financial decision-making is the primary mechanism of how our societies balance the conflicting demands on how we use those monetised and limited resources.

Financiers are the stewards of our societies financial resources that have been aggregated by bank savings, pensions, insurance funds, wealth funds, family offices, high net worth individuals, etc. These financial resources have become the monetary measure of our 'Wealth of Nations' that was envisioned by Adam Smith. And as such, we rely on financiers to invest our societies' capital to achieve optimum returns.

However, the space is mired in conflicting interests that are of importance to sustainability, for example: Financiers' performance metrics and bonuses are normally based on short-term financial metrics of only a few years duration, yet the goals of financiers' clients are often wealth creation over decades and even transcending family generations.

Yet capital investment in infrastructure assets requires a long-term technical view, to plan for a long-term low impact on the environment and society, to plan for long-term utility, to have resilience to changing consumer behaviours, to be resilient to external impacts such as how climate change might impact performance, etc. Moreover, a persistent challenge for financiers of infrastructure projects is the dichotomy of short-term financial planning cycles versus the need to amortise costs over an extended period to make infrastructure affordable.

Irrespectively the clients of financiers expect financial returns on the monies invested, and unless sustainability attributes have been articulated in clear financial terms, generally financiers have seen sustainability at best as a tick-box compliance exercise, rather than a core component of investment planning. Fortunately, this landscape is rapidly changing.

This chapter explores how financiers consider sustainability in infrastructure financial decision-making, and how the infrastructure industry can effectively communicate on sustainability so that financiers are appropriately informed in financial decision-making to deliver infrastructure that will sustainably provide utility over decades and even centuries, by exploring the following three questions:

1. **Do financiers know how sustainable are current business practices?** We will explore how current financial reporting influences decision-making, and also the exposure arising from sustainability disruption not being sufficiently incorporated into that decision-making.
2. **How can we mainstream sustainability into our future business practices and into financiers' decision-making on resource allocation?** We will explore disruptions, sustainable resource-use planning, sustainability metrics

and reporting and how these practices can inform decision-making to drive more resilient business.

3. **How can we encourage financiers to focus on longer-term sustainability goals rather than short-term financial gain?** We will explore frameworks for dynamic monitoring and reporting of sustainability metrics and risks and how the market is moving to using these metrics so that appropriate financing is committed throughout the entire asset lifecycle to continuously provide sustainable outcomes.

Whilst this chapter has been primarily written to describe emerging best practices in the private sector in communicating on sustainability for financial decision-making, the observations provided herein are equally applicable to public sector in appraising the long-term socio-economic and environmental cost-benefits of sustainable infrastructure.

10.3 DO FINANCIERS KNOW THE SUSTAINABILITY OF CURRENT BUSINESS PRACTICES?

Conventional financial performance reporting is, by definition, retrospective, yet it is the main instrument by which financiers make decisions in planning and controlling business. However, conventional financial performance reporting does not adequately capture sustainability risks and opportunities, leaving decision-makers blind to the winds of change being brought by sustainability disruption. Financial decision-making needs to be fully informed of both sustainability-linked upside opportunities and downside risks.

Conventional boardroom discussions and decision-making are predominantly based around a retrospective review of financial performance reporting, namely the all-to-familiar last 3 years of revenues, expenditures, assets and liabilities accompanied by statements from the Chief Executive Officer (CEO), Chief Financial Officer, Chief Operations Officer, etc. (collectively 'the C-suite') on historic performance, market conditions and business risks (Table 10.1).

Similarly, C-suite statements on forecasts of future revenues and costs normally use historical data to project forward those historical trends, with adjustments for parameters such as inflation, changes in GDP or population changes. Also, these financial reports are relied upon by financiers in appraising asset valuations when buying/selling

TABLE 10.1 Conventional financial performance reporting

CASHFLOW	*BALANCE SHEET*
Income	Asset
Expenditure	Liability
Risks and opportunities	

a business or negotiating financing. This is also the case for most infrastructure investments where historical consumption patterns and their growth are the base case for projecting future demand.

However, there are at least three key challenges with the conventional financial reporting approach that leave decision-makers blind to the winds of change:

- **It is based on a retrospective dataset**: Conducting our financial resource allocation based on historic financial and economic data, then projecting forward these historic trends, means that financial decision-making is done without meaningful inclusion of the realities and risks of sustainability disruptions.
- **It takes a short-term view of business**: By reporting only the last few years of financial performance and typically only projecting forward a few years deprives decision-makers of insights from longer-term datasets, where the trends and also disruptions therein within our economies and environment would be more apparent. There is often insufficient consideration, costing and budgeting of future costs such as asset renewals or decommissioning costs in projections.
- **It does not internalise the full costs of doing business**: Many businesses have become decoupled from both the communities and the environment in which they do business; thus, the costs of goods and services at the point of purchase are priced through market dynamics at the financial cost, not the real and full economic cost of doing business. Governments are laggards in bringing regulations to require businesses to address adverse socio-economic and environmental impacts of their business. Market leaders recognise that their corporate actions have wider implications and have embraced their Corporate Social Responsibilities, and/or taken on Extended Producer Responsibilities to incorporate socio-economic factors into decision-making and mitigate impacts, yet such practices need to be more widely adopted across all of industry.

These short-sighted practices fundamentally mean that financiers are not optimally allocating a nation's limited resources for their society's long-term future and consequently undermine the long-term resilience of a nation's business enterprises. In the infrastructure space, such short-sighted practices equate to undermining the resilience of communities, economies and environment to be able to mitigate impacts from sustainability disruptions. Moreover, for owners and financiers, such an approach can also woefully overvalue business valuations by misrepresenting potential future revenues and future real costs of doing business.

Reflecting on the above in philosophical terms, Adam Smith deliberately published his *Theory of Moral Sentiments* (Smith, 1812) to promote awareness of moral values as the basis of decision-making before publishing *The Wealth of Nations*. And indeed, we find that our contemporary financial decision-making frameworks are weak in protecting values necessary for promoting the 'Wealth of Nations' and certainly do not provide the basis for guiding our economies to a sustainable future that can provide social equity.

Practical examples of this divergence from sustainable planning and investment include the following:

1. Most investment in the power sector is attracted to new generation capacity, despite there being adequate redundant capacity in most power networks, and this bias is to the detriment of grid upgrades (even though grid investments typically on long-term provide stable returns) and demand management (it is harder to sell a business case based on reducing consumption/reduced gross revenues, even when it is based on more efficient resource use).
2. Even with modern green building rating systems, the construction of real estate still does little to internalise the full environmental and economic costs of construction (e.g. the environmental impacts of manufacturing steel and concrete, mitigating high energy operating costs and planning for end-of-life demolition).
3. Development Banks still largely build the economic business case for transport projects on vehicle operating cost savings and time cost savings, plus accident cost savings, and reduction in environmental impacts (ADB, 2017). However, a values-driven approach would recognise and build the economic business case to be on creating socio-economic connectivity, and that mobility is just one of the key services required in creating a vibrant socio-economic corridor. This approach will drive the design brief to meet a wider range of needs and will likely result in higher asset utilisation and a more favourable financial performance.
4. Demand assessments for utilities in development contexts still typically are based on interviews with the head of the household and are based on consumption patterns. Widening the demand assessments to specifically differentiate on gender-based perspectives of what is the productive use of those utilities would typically reveal a differentiated perspective on real tangible value and the corresponding willingness-to-pay for services.

Instead, we need to invoke a term, coined in the 1860s by another philosopher John Ruskin, 'illth' (Ruskin, 1910) being that part of a nation's natural and/or monetary wealth being inefficiently managed to the detriment of society and the environment. We need to be challenging our financiers to prevent illth and instead hold them accountable in demonstrating that they are using appropriate values to create financial value that also has socio-economic value is thus sustainable wealth for our nations.

Financial reporting systems therefore need to bring a richer and more relevant dataset that can be used to appraise the performance of our financiers in allocating our limited resources to create social wealth or illth, with remuneration and bonuses structured accordingly.

We need to consider how financial reporting procedures and practices are evolving and how we can better communicate to understand how we need to better inform boardrooms of the real needs of our communities, economies and environment. A quick review of the current perspectives of measuring corporate performance reveals the breadth and depth of vulnerability in our current business practices:

Income: Historically, the incomes of infrastructure services have been disrupted by technology innovations (see Box 10.1 for examples) that have changed volumes and characteristics of demand. These traditional market disruptions had corresponding impacts on incomes such as loss of market share, loss of revenues and erosion of margins.

We are now seeing incomes of infrastructure services being disrupted by sustainability-related issues that change the volume, pricing and other characteristics of demand and therefore income. Such sustainability disruptions include shifting consumer perceptions, increasing price of goods and services, affordability and willingness-to-pay, and diminished trust in brand from disclosure of business practices that are not in keeping with emerging trends.

Expenditure: Sustainability disruptions can quickly cause cost escalations that erode business margins; examples are given as follows:

- Higher business risk profile increasing cost of finance and insurance
- Higher taxes/carbon taxes/carbon credits/waste disposal taxes
- Higher capital costs to meet more stringent standards, for example, internalising the real costs of steel or cement manufacture, have significant implications for the infrastructure sector
- Increased costs of raw materials; especially our exponential increase in resource consumption which has led to depletion of readily extractable virgin resources, leading to increase in raw material costs
- Increased costs of land in securing more resilient sites for new operations or in increasing resilience of existing operating sites
- Increased labour costs
- Increased energy and utility costs

Assets: They are typically resources owned/controlled by a company that for accounting purposes are quantified and reported in terms of their realisable value, such as cash, land, buildings, facilities, plant and equipment, accounts receivable, intellectual property and goodwill.

The International Financial Reporting Standards (IFRS) has various standards for various instruments. Standard 9 on Financial Instruments was updated in 2018, including requiring that all foreseeable impairments on assets are reported in financial reporting and are incorporated into asset valuations (IFRS, 2024). Whilst sustainability-linked risks are not explicitly defined in the IFRS, it is now widely accepted financial auditing practice that sustainability-linked risks are often systemic and therefore should be incorporated into reporting (but often are not due to a myriad of reasons from ignorance, or not being sufficiently robustly communicated in monetary terms, or from wilful negligence).

Emerging sustainability-related issues that impair asset valuations include the following:

1. How resilient are the business's land, buildings, facilities, plant and equipment to sustainability disruptions? Are there sustainability disruptions that impact condition of such facilities and thus impair asset values?

BOX 10.1: EXAMPLES OF INFRASTRUCTURE SECTOR EVOLUTION THAT HAVE DISRUPTED MARKETS

The evolution of communication infrastructure: postal services; the telegraph; conventional landline telephones; the fax machine; email, Blackberry, 2G, 3G, 4G and 5G connectivity, mobile and smart phones; and internet communication.

The evolution of energy services infrastructure: biomass, charcoal, coal, natural gas, nuclear, and renewables.

The evolution of cargo transport sector: mules, canals, steam locomotive, diesel locomotive, electric locomotive, container ships and post-Panamax container ships.

2. Have measures been taken to identify, quantify, report and minimise such exposures and have provisions been made?
3. Where customers' businesses have been hit by sustainability disruptions (e.g. factories hit by flooding), are they still able to honour their accounts due (the money they owe)? How well is such business exposure and value-chain system risks understood?
4. How resilient is the business to policy changes or changes in market demand that could result in assets becoming stranded (e.g. Germany's decision to close its nuclear power plants, or the shift to close coal power plants before their natural end of asset life)?
5. Is the valuation of goodwill[1] a fair representation of the business within the dynamics of sustainability disruption, especially reputational risk?

Liabilities: Liabilities are typically credit lines, loans, bonds and accounts payable. Emerging sustainability-related liabilities include the following:

1. Lenders have increasingly demanding compliance requirements such as through their pledges to comply with the Paris Agreement (aka Paris Accords) that stymies financing of coal-fired power assets, *inter alia*, and generally are seeking considerably more data from infrastructure sector borrowers on the broader socio-economic and environmental impacts of their business activities. Ostensibly, borrowers that can show financiers a lower risk profile should be able to negotiate more favourable financing.
2. Levies imposed by governments for failure to meet climate-related targets or to discourage or encourage specific activities.
3. Insurance on infrastructure normally comes with an obligation on the Insured Party to notify the insurer of material changes in risk profile. However, the short-term views taken in business and the cumulative incremental nature of sustainability disruptions often leave such change in risk profiles going unnoticed/unreported. An insurer would not be obliged to pay out where the Insured Party has failed to disclose such exposures. In more extreme cases such as observed in some mining and coal power projects, sustainability

disruptions have resulted in withdrawal of insurance, resulting in uncovered business risks. Insurance is also often required as covenants to loans; thus, loss of insurance can also trigger a need to restructure debts.

4. **Contingent liabilities:** A key challenge in financial performance reporting of infrastructure assets is a general weakness in reporting contingent liabilities, i.e. potential expenditures that may occur in the future, but are unbudgeted and potentially uninsurable. Conventionally, these may be pending lawsuits or honouring warranties but increasingly could also be unbudgeted expenditures to mitigate sustainability-related or climate-related impacts such as loss of business due to a storm event, unexpected repairs due to flooding, relocating critical infrastructure, and building flood defences, or the withdrawal of insurance coverage as the insurance market deems certain risks as uninsurable.

Risks and opportunities: We can see that sustainability disruptions have become material in impacting the financial performance of infrastructure assets. With infrastructure investments having 'price tickets' from tens of millions to billions of dollars, often with long development and construction periods, financiers are understandably nervous about risks that could diminish returns on their investments. Boardrooms are now increasingly aware that business risks arising from sustainability-related disruption can have material impacts on financial performance with topics listed below now regularly part of C-suite conversations. Examples of sustainability-related disruption risks impacting financial performance are tabulated in Table 10.2.

Thus, it is evident that for businesses to be part of a resilient future that can mitigate the worst of sustainability disruptions will require deep change in the frameworks within which we measure, report and make decisions on our resource allocation.

In particular, we need our socio-economic planning and our financial decision-making to be guided by data-driven, comprehensive assessments which are set in the context of long-term trends and cognisant of risks from sustainability disruptions. This planning should seek to address the real needs of our communities, economies and environment whilst also incorporate the real costs of sustainable service provision so that boardrooms and financiers deliver more sustainable resource allocation.

10.4 HOW CAN WE MAINSTREAM SUSTAINABILITY INTO OUR FUTURE BUSINESS PRACTICES AND FINANCIERS' DECISION-MAKING ON RESOURCE ALLOCATION?

Sustainability has become recognised in boardrooms as a significant business disruptor. New sustainability goals and frameworks have been developed, which enable consistent and structured conversations on sustainability disruptions, setting shared goals

TABLE 10.2 Sustainability-related disruption risks impacting financial performance

RISK CATEGORIES	*EXAMPLES OF SUSTAINABILITY-RELATED DISRUPTION RISKS IMPACTING FINANCIAL PERFORMANCE*
Political risks	Citizens, voters, NGOs and business owners are feeling the consequences of sustainability disruptions on their livelihoods and assets and pressing governments to change policies in responses to climate change, energy transition, urbanisation, provision of transport infrastructure, water resource management, etc. These policy changes can have fundamental impacts on the demand for infrastructure and change the obligations and costs of doing business for infrastructure asset owners. The 'great divergence' of wealth and growing inequality continues to erode social harmony and threatens further political turmoil.
Market risks	The demand for infrastructure has continually evolved; trends are normally more obvious when appraised over decades but can also happen in relatively short periods. The financial returns on investments in infrastructure assets are very sensitive to changes in end-user demand, so financiers will normally appraise all foreseeable disruptors to demand. But such sensitivity assessments are normally based on historic data and historic events, so do not provide a comprehensive assessment of sustainability-related disruptions. There is growing awareness that we have had an exponential growth in consumption as shown in Chapter 1 with large population expansion and economic per capita growth, resulting in both localised and global depletion of resources with corresponding shortages and polarisation in access to resources. Moreover, innovation brings new technologies that disrupt traditional supply-side solutions for serving demand.
Insurance risks	The insurance market sentiment has shifted significantly in certain asset classes due to perceptions of exposure to systemic risks in sustainability-related disruptions such as the observed impact of the Paris Accord (UNFCCC, 2015) on the coal-fired power industry and the mining sector and increasingly so for assets such as real-estate and industrial estates prone to flood risk (International Actuarial Association, 2019). Especially, insurers of such assets are finding it progressively more expensive or difficult to on-sell and dilute their risks in the reinsurance markets.[a]
Physical risks	Whilst flooding has been one of the most visible and significant sustainability-related impacts on infrastructure assets, there are numerous physical risks arising from changes in our physical environment (e.g. seawater level rises, storm events, coastal erosion, droughts and water shortages, changes in ambient temperatures, etc.) that diminish infrastructure performance or even impair the condition and value of infrastructure assets.

[a] Reinsurance is "insurance for insurance companies": insurance companies spread their financial risks through buying reinsurance and so dilute the impact of catastrophic loss events.

and shared metrics for monitoring and reporting performance. These are powerful in helping governments, business and communities to align to shared sustainability goals.

Until recently, financiers have not had sufficient incentive or information available to analyse the financial or economic costs and opportunities from sustainability disruptions; thus, sustainability has largely been a side conversation in financial decision-making.

However, in recent years governments have been under considerable social pressure to make progress in ratifying and adopting a range of sustainability goals and frameworks which also require boardrooms and financiers to consider sustainability. What is particularly helpful is that these sustainability frameworks enable consistent and structured conversations on sustainability disruptions, setting shared goals and also shared metrics for monitoring and reporting performance. Some significant examples are listed below together with how these are disrupting traditional business and especially provide guiderails for those businesses making smart pivots toward more sustainable business models (Table 10.3).

Aligning with the UN Sustainable Development Goals (SDGs) is good business: From a business perspective, the SDGs are particularly powerful in challenging conventional thinking, as these 17 goals effectively cover all the core needs of our communities. For example, communities will always have demand for businesses that deliver water resource management services, health services, education services, energy services, communications services, etc. Moreover, each of the SDGs are accompanied by metrics of our communities' fundamental needs and reveal the trends which give decision-makers greater visibility and insights on where our limited resources need to be prioritised. Businesses that have made smart pivots orientated toward addressing these fundamental socio-economic needs can expect to be both more resilient and intrinsically more sustainable. Therefore, in planning and providing infrastructure services, it is prudent for infrastructure asset developers and owners to endeavour to incorporate all relevant SDGs and their associated metrics into long-term performance-based contracts. Many businesses have already reviewed the SDGs and the associated metrics to determine both how well aligned their business activities are to the goals and also to appraise their exposure where there is misalignment.

However, these goals and frameworks are not enough to drive our objective of long-term financing commitments throughout the entire asset lifecycle to continuously deliver sustainable outcomes. To encourage financiers to mainstream this objective requires that technical experts appropriately inform financiers of how sustainability disruption impacts the costs of financing.

10.4.1 Measuring Costs of Sustainability Disruption in the Financing

Sustainability Reporting Frameworks and Standards provide a structured and consistent way of bringing relevant information to assist financiers to understand exposure to sustainability disruption. Moreover, Sustainability Ratings and Indices help financiers

TABLE 10.3 Sustainability goals and frameworks providing guiderails making smart sustainable pivots

SUSTAINABILITY GOALS AND FRAMEWORKS	*WHAT ARE THESE SUSTAINABILITY GOALS/ FRAMEWORKS?*	*HOW DO THESE SUSTAINABILITY GOALS/ FRAMEWORKS DISRUPT TRADITIONAL INFRASTRUCTURE BUSINESS?*
United Nations (UN) Global Compact[a]	An international framework for business to actively engage in sustainable and socially responsible policies in a structured framework comprising ten core principles, and obligations on reporting on implementation.	Businesses can expect that engagement will give a competitive market advantage through demonstrating leadership by advancing responsible corporate citizenship; producing practical solutions to contemporary problems related to globalisation, sustainable development and corporate responsibility in a multi-stakeholder context; managing risks by taking a proactive stance on critical issues; leveraging the UN's global reach and convening power with governments, business, civil society and other stakeholders; sharing good practices and learnings; accessing the UN's broad knowledge in development issues; improving corporate/brand management, employee morale and productivity, and operational efficiencies.
UN Sustainable Development Goals (SDGs) (United Nations, 2015).	The UN SDGs are a set of 17 goals adopted by all UN Member States in 2015 as a universal call to action to end poverty, protect the planet and ensure that by 2030, all people enjoy peace and prosperity. The goals are a call for action by all countries – poor, rich and middle-income – to promote prosperity while protecting the environment.	Move governments' goals and expectations for infrastructure, especially driving a shift in how infrastructure-based services are delivered. Consequences for infrastructure business include the following: Governments are modifying market conditions to favour businesses that are aligned to delivering the SDGs. Governments need to regularly report on their progress toward achieving the indicators as provided in the SDGs and so are putting obligations on businesses to also align their reporting with SDG metrics (United Nations, 2015). These sustainability disruptions create opportunities for businesses that are dynamic in applying innovation, reducing risks and costs and building stronger relationships with customers and stakeholders.

(*Continued*)

TABLE 10.3 (*Continued*) Sustainability goals and frameworks providing guiderails making smart sustainable pivots

SUSTAINABILITY GOALS AND FRAMEWORKS	*WHAT ARE THESE SUSTAINABILITY GOALS/ FRAMEWORKS?*	*HOW DO THESE SUSTAINABILITY GOALS/ FRAMEWORKS DISRUPT TRADITIONAL INFRASTRUCTURE BUSINESS?*
The Paris Agreement and Nationally Determined Contributions (NDCs)	The Paris Agreement (aka Paris Accords) is a legally binding international treaty on climate change (UNFCC, 2015). It was adopted by 196 Parties (i.e. countries) at the UN Climate Change Conference (COP21) in Paris, France, on 12 December 2015. The Agreement aims to limit global warming to well below 2°C above pre-industrial levels and pursue efforts to limit the temperature increase to 1.5°C above pre-industrial levels. Each Party to the Paris Agreement is required to establish 'Nationally Determined Contributions' as part of their commitment to the Paris Agreement. NDCs are non-binding national plans for mitigating and adapting to climate change. The NDCs must include national targets for greenhouse gas emission reductions, policies and measures that governments plan to implement in response to climate change and as a contribution to achieve the global targets set out in the Paris Agreement.	For governments to achieve their commitments under the Paris Agreement, there are significant changes being implemented. Governments are implementing wide reaching policy revision, which include setting climate mitigation and adaptation targets (NDCs) across a range of relevant sectors and carrying these into laws and regulations to assign clear obligations. Governments are implementing wide-reaching reviews and revisions to subsidy frameworks, tax frameworks, to provide a motivating environment. Governments are implementing wide-reaching reviews and revisions of planning, implementing and monitoring frameworks to mainstream climate mitigation and adaptation action and continuous improvement. Governments are passing down obligations for Regionally and Locally Determined Contributions and Voluntary Local Review (Collaborative Climate Action, 2021).

[a] *See:* https://unglobalcompact.org/.

quantify gaps and shortfalls in a business's sustainability performance relative to peers and so enable them to make informed decisions on sustainability risks and the corresponding costs of financing.

The good news is that in recent years, there has been an explosion in Sustainability Reporting Frameworks, Sustainability Reporting Standards, Sustainability Ratings, Sustainability Indices, Sustainability Taxonomies, etc. that can assist financiers understand exposure to sustainability disruption and to make informed decisions. However, this plethora of sustainability financial reporting tools have also caused much confusion on which of these should be used and when. We will define these sustainability financial reporting tools and consider their functions before considering how we can best interact with these tools to bring relevant information to decision-makers.

10.4.2 The Spreading Adoption of Sustainability Reporting by Financing Institutions

Mark Carney in his time as Governor of the Bank of England was a leading advocate in encouraging businesses to recognise that sustainability disruptions can have both material and systemic disruption to financial performance, and also, that these sustainability disruptions should be disclosed in financial reporting to be a core part of C-suite business decision-making for enabling long-term resilience of businesses with a corresponding positive influence on economies. The underlying rationale was that if financing institutions recognised the potential material and system risks to their financial performance, then these financing institutions would require those businesses that they were financing, investing in, lending to and insuring to also monitor, report and plan to mitigate on sustainability-related impacts and risks, thus achieving accelerated widespread adoption.

10.4.3 Jargon Busting

Sustainability Reporting Frameworks provide organisations – large or small, private or public – with guidelines on how to report on their sustainability performance. These frameworks provide a common language and structure for companies to report on their resource use and environmental, social and governance (ESG) performance. Sustainability Reporting Frameworks are normally accompanied by Sustainability Reporting Standards.

Sustainability Reporting Standards provide guidance on disclosure of financially material sustainability information to financiers in an industry-based and comparable way. Thus financiers – across markets, asset classes and strategies – can integrate sustainability considerations into their investment decisions.

Sustainability Taxonomies are classification systems that define which economic activities are sustainable, normally from an ESG perspective. Such taxonomies are used to help investors characterise the potential sustainability of investments and communicate in a consistent way within the business community to achieve common understandings.

Sustainability Ratings are usually generated by third-party organisations for use by investors to obtain an independent and objective assessment of an organisation's sustainability performance relative to a recognised Sustainability Reporting Standard and communicated using Sustainability Taxonomies so that the ratings are industry-based and comparable.

Sustainability Indices are similarly used to provide independent and objective assessment of sustainability performance but are primarily applied to companies listed in stock market indices that track the performance of companies that meet certain sustainability criteria.

Critically, Sustainability Reporting Standards have become hurdles that governments, businesses and projects must clear to be able to access favourable financing from institutions that have subscribed to Sustainability Reporting Frameworks and Standards. Sustainability Ratings and Indices have become an essential tool for financiers to assess which businesses to finance and for pricing the risks and therefore the cost of that financing.

10.4.4 How Sustainability Reporting Standards and Indices Impact the Infrastructure Business

Examples of these sustainability reporting tools and how these are influencing infrastructure businesses as guiderails for pivoting toward sustainable business solutions are shown in Table 10.4.

In drawing initial conclusions from this review, financiers are now using these Sustainability Reporting Frameworks and Tools to inform their appraisals of which businesses to finance and for pricing the risks and therefore the cost of that financing. There is an underlying shift from the principle of 'do no harm', which is a compliance-based approach, to the principles of 'do better', which is an approach striving for continuous improvement through proactive engagement in sustainable and responsible business practices to create greater socio-economic value and more resilient business.

Thus, the challenge for planners, engineers, technologists and scientists of infrastructure is to understand how find effective ways to proactively feed relevant information to enable informed data-driven appraisals of solutions that 'do better'.

10.4.5 How Can We Keep Financiers Focused on Longer-Term Sustainability Goals Rather Than Short-Term Financial Gain?

We can engage in challenging conversations with financiers on the vulnerabilities of conventional approaches which intrinsically focus on value preservation, which is simply staying in business for now and move the conversation to progressive 'value creation and capture' approaches for sustainably growing business.

TABLE 10.4 Sustainability reporting tools and their influence on infrastructure businesses

SUSTAINABILITY FINANCIAL REPORTING TOOL	*WHAT ARE THEY?*	*HOW ARE THEY INFLUENCING INFRASTRUCTURE BUSINESSES?*
World Bank Environmental and Social Reporting Framework (World Bank, 2016)	The WB's ESF is a Sustainability Reporting Framework that comprises A Vision for Sustainable Development; Ten Environmental and Social Standards, which set out the requirements that apply to Borrowers; An Environmental and Social Policy for Investment Project Financing (IPF), which sets out the requirements that apply to the Bank; An Environmental and Social Directive for IPF; and A Directive on Addressing Risks and Impacts on Disadvantaged or Vulnerable Individuals or Groups.	Governments and businesses that are seeking to use World Bank financial products and instruments on their projects and enjoy the corresponding benefits must demonstrate the Project's compliance to the WB's ESF requirements. Consequently, the impacts for infrastructure projects include strengthening environmental and social management systems and institutions, and Borrower capacity building. Projects require enhanced transparency and stakeholder engagement through timely information disclosure, meaningful and ongoing consultations are required throughout the project lifecycle, and responsive grievance mechanisms are required to facilitate resolution of concerns and grievances of project-affected parties.
IFC's Environmental and Social (E&S) Performance Standards (IFC, 2012)	As part of IFC's Sustainability Framework, Policies and Standards, the IFC's E&S Performance Standards are a Sustainability Reporting Framework that define IFC clients' responsibilities for managing their environmental and social risks.	IFC's E&S Performance Standards have become the global industry benchmark standard for identifying, understanding, assessing and managing environmental and social risks when financing Projects. Consequently, compliance to IFC E&S PS from the project preparation phase can reduce risk of project delays and increase likelihood of all parties to the agreements in recognising the importance of mainstreaming E&S compliance.

(*Continued*)

TABLE 10.4 (*Continued*) Sustainability reporting tools and their influence on infrastructure businesses

SUSTAINABILITY FINANCIAL REPORTING TOOL	*WHAT ARE THEY?*	*HOW ARE THEY INFLUENCING INFRASTRUCTURE BUSINESSES?*
The CEO Water Mandate (UN, 2007)	The CEO Water Mandate is a Sustainability Reporting Framework. This is a UN Global Compact initiative that seeks to mobilise business leaders to advance corporate water stewardship as part of the SDGs. Endorsers of the CEO Water Mandate commit to continuous progress against six core elements of water resource stewardship and in so doing understand and manage their own water-related business risks.	The CEO Water Mandate aims to increase awareness amongst stakeholders of the impacts that corporations have on water resources. Corporations that adopt the CEO Water Mandate can expect to benefit from other members sharing best and emerging practices, be better able to demonstrate active engagement in water resource stewardship, and be better able to maintain access to water resources that are essential for conducting their business activities.
Equator Principles (Equator-Principles, 2020)	The Equator Principles are a Sustainability Reporting Framework for reporting to lenders (especially commercial lenders) on environmental and social risks arising in infrastructure and industry projects that use limited or non-recourse project finance. The principles are based on the International Finance Corporation's (IFC) Performance Standards on Environmental and Social Sustainability.	The Equator Principles set out minimum requirements for Project Developers seeking to borrow on a 'Project Finance' basis from financial institutions. The Equator Principles have been instrumental in informing Lenders of key environmental and social risks associated by giving a standard framework for identifying, understanding, assessing and managing environmental and social risks when financing Projects. The principles also help financial institutions meet the expectations of stakeholders who are concerned about environmental and social issues.

(*Continued*)

SUSTAINABILITY FINANCIAL REPORTING TOOL	*WHAT ARE THEY?*	*HOW ARE THEY INFLUENCING INFRASTRUCTURE BUSINESSES?*
Task Force on Climate-Related Financial Disclosures (TCFD, 2019)	The TCFD is a Sustainable Reporting Framework that provides recommendations on the types of information that companies should disclose to support investors, lenders and insurance underwriters in appropriately assessing and pricing risks related to climate change.	Climate change presents financial risk to the global economy. Financial markets need clear, comprehensive, high-quality information on the impacts of climate change. This includes the risks and opportunities presented by rising temperatures, climate-related policy and emerging technologies in our changing world. The TCFD has a deep impact on how an organisation conducts business, from how it carries out risk management through to disclosures on their exposure to climate-related risks and opportunities, which forces increased transparency and comparability of climate-related financial information.
Global Reporting Initiative (GRI, 2013)	The GRI Standards are Sustainability Reporting Standards designed to enable any organisation – large or small, private or public – to understand and report on their impacts on the economy, environment and people in a comparable and credible way, thereby increasing transparency on their contribution to sustainable development. In addition to companies, the Standards are highly relevant to many stakeholders – including investors, policymakers, capital markets and civil society. The Standards are a modular system comprised of three series of Standards to be used together: Universal Standards, Sector Standards and Topic Standards.	For organisations to comply to the GRI standards would require business policies to be aligned to the GRI standards. Organisations may find that they need to monitor and report on more parameters than in business as normal. Organisations can use the GRI Standards to prepare a sustainability report in accordance with the Standards or use selected relevant Standards (or relevant parts of their content) to disclose information for specific users or purposes, for example, climate change impacts for investors and consumers. Organisations may wish to engage an independent third-party verifier of their sustainability reports to provide assurance to stakeholders that the information is accurate and also demonstrate the organisation's commitment to accountability and transparency.
Sustainability Accounting Standards Board (SASB) Standards (SASB, 2019).	SASB Standards are Sustainability Reporting Standards, for communicating financially material sustainability information to investors in an industry-based, metric-driven and comparable way to enable integration of sustainability considerations into investment and stewardship decisions across global portfolios and asset classes.	SASB Standards identify the subset of environmental, social and governance issues. The SASB requires that organisations follow a prescribed technical protocol for compiling data to be able to provide investors with comparable data that feeds the financial data and analytics ecosystem. This may impact on how organisations monitor and report information.

(*Continued*)

TABLE 10.4 (*Continued*) Sustainability reporting tools and their influence on infrastructure businesses

SUSTAINABILITY FINANCIAL REPORTING TOOL	*WHAT ARE THEY?*	*HOW ARE THEY INFLUENCING INFRASTRUCTURE BUSINESSES?*
MSCI (2022), Sustainalytics (2024)	Morgan Stanley Capital International and Sustainalytics are examples of Sustainability Ratings providers to institutional investors and companies. These Ratings Agencies use rules-based methodologies to identify industry leaders and laggards according to their exposure to ESG risks and how well they manage those risks relative to peers.	These ratings agencies receive their income from institutional investors and financiers that are seeking objective and independent data to assist financial decision-making. Consequently, these Ratings Agencies will seek to validate the ESG disclosures made by companies through independent research ranging from monitoring government data sources, Regulatory Agency data, monitoring adverse media reports, third-party data sources, etc. The Ratings Agencies will also benchmark performances on a wide variety of metrics. These independent appraisals and industry benchmarking help to challenge companies to 'keep reporting honest', prevent greenwashing, and maintain pressure for companies to continuously improve to maintain competitivity in the financing markets.
FTSE Russell ESG (2024), Dow Jones Sustainability World Index	The FTSE Russell ESG Index and Dow Jones Sustainability World Index are examples of Sustainability Indices. These were designed to help investors align investment and ESG objectives while maintaining industry neutrality. The ESG Ratings and data model cover securities (especially stocks and bonds) and measures corporate exposure to critical ESG issues, assessing operational risks and performance drivers. The ESG Ratings are made up of several levels of data: one overall rating, followed by thematic assessments, with each theme having multiple quantified indicators, offering an objective and flexible approach to integrating ESG considerations into investments and stewardship.	Financiers use this data as the basis for active portfolio management, company engagement, customised indexes, ESG risk analysis, or research. These independent appraisals and industry benchmarking help to challenge companies to "keep reporting honestly, prevent greenwashing, and maintain pressure on companies to continuously improve to encourage competitiveness in the financing markets".

Planners, engineers, technologists and scientists of infrastructure need to transform how they engage in conversations with financiers to formulate compelling propositions of what sustainability really looks like.

10.4.6 The Conventional 'Value Preservation' Approach – Simply Staying in Business for Now

The conventional approach that financiers and business owners take to sustainability can be characterised as 'value preservation', i.e. simply acknowledging the most recent changes to incomes, expenditures, impacts on assets and liabilities and the most pressing risks and opportunities with an outcome of simply 'how do we stay in business for now?'

10.4.7 The Progressive 'Value Creation and Capture' Approach – Sustainably Growing Business

In identifying, planning, procuring, constructing, operating and maintaining truly sustainable infrastructure services, we need to challenge conventional and narrow

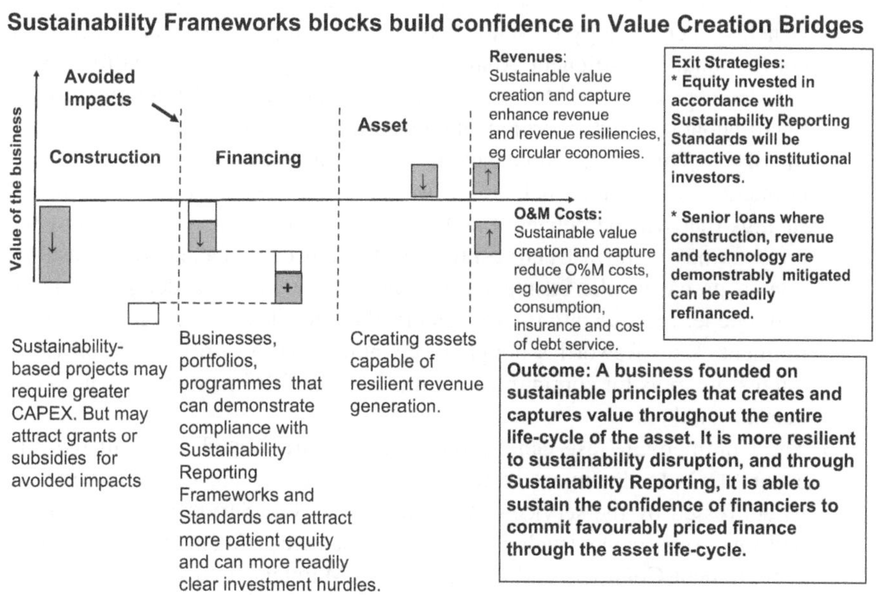

FIGURE 10.1 Illustration of how Sustainability Frameworks provide blocks for building confidence in Value Creation Bridges.

approaches by recognising the full spectrum of impact and benefits that infrastructure provides, then identifying ways for smart pivots **to create and capture value throughout the entire lifecycle of an asset with diversified resilient revenue streams; thus, the outcome is sustainably growing business** (Figure 10.1).

We need to communicate how mainstreaming sustainability can provide value creation and capture opportunities. This communication needs to use formats, narrative and vocabulary that can be readily understood by financiers and tested in their Financial Models when appraising financing options. An example is a 'Value Creation Bridge' that can illustrate how actions can prevent value loss and preserve value in times of disruption, while also strengthening the long-term value of the business.

10.4.8 How to Implement a Progressive ('Value Creation and Capture') Approach

The starting point is using guidance such as the Better Business Case framework (HM Treasury, 2018) to develop a comprehensive value proposition that gives compelling arguments and information to financiers, which comprises the following five business case components for an infrastructure service:

1. The Strategic Case can be articulated using all the relevant SDGs to explain which essential needs of our communities will be addressed by the services.
2. The Economic Case can leverage guidance on identifying and quantifying the 'Wider Economic Benefits'[2] that will be delivered. The 'Wider Economic Benefits' methodology of analysis is a more comprehensive approach than conventional Cost–Benefit Analysis by actively planning for indirect benefits from projects (Australian Transport Assessment and Planning, 2022). Socio-economic and environmental performance metrics can be drawn from the SDGs. This enables financiers to have informed conversations with governments on the additional social and economic benefits, fiscal revenues or savings that will be delivered and can strengthen public sector support for programmes of infrastructure projects. Having deep stakeholder consultation in preparing the business case can provide financial decision-makers with deeper insights into the real and resilient needs of society and the environment. Kate Raworth's doughnut economics (Rayworth, 2017) provides interesting tools for communicating the depth of fulfilment of these needs to decision-makers.
3. The Commercial Case can similarly also be articulated using all the relevant SDGs to document which essential needs of our communities will be addressed by the services. Performance of 'The Services' can be specified into Agreements (such as Service Agreements, Financing Agreements, Procurement Contracts and Operation and Maintenance Contracts) by also drawing from the socio-economic and environmental metrics listed under each of the relevant SDGs. Agreements can also oblige the Parties to create a common digital data environment (see more detail below) that provides

instant access and reporting performance of all the key metrics required by the Sustainability Reporting Frameworks and Sustainability Reporting Standards, thus bringing relevant information to financiers in a readily useable format, whilst also building both transparency and trust. A key differentiator is that the progressive approach is more likely to identify additional revenue streams and cost savings that can significantly strengthen the financial case.

4. A key component of building a strong commercial case is to have procurement strategies that enable the business to pass down sustainability obligations to its suppliers, including having back-to-back reporting obligations passed down. This has become a mature practice in European waste management practices through regulated Chain-of-Custody obligations as implemented through mechanisms such as waste transfer notes (Gov UK, 2021). In the climate change space, we are now seeing the emergence of Scope 1, 2 and 3 reporting obligations on the embedded greenhouse gas emissions in goods and services (UNFCC, 2023). Simply, Scope 1 is the quantity of direct emissions your organisation releases in production, Scope 2 emissions are indirect emissions from the generation of energy purchased by your organisation and Scope 3 emissions are indirect emissions that occur in the value chain of an organisation, both upstream and downstream.
5. The Financial Case when it has been developed through the above process is then based on sustainable principles and so will likely be more resilient to changes in socio-economic and environmental factors making revenues and costs also more resilient. Assets will be less vulnerable to impairments. And liabilities, including contingent liabilities, should be better understood and more readily mitigated. The business will be aligned with appropriate Sustainability Frameworks and Sustainable Reporting Standards that can be readily understood by financiers. Consequently, the business should be a more compelling proposition for financiers and should have better access to more favourably priced financing.
6. The Management Case when it has been developed through the above process will demonstrate that robust arrangements are in place for the delivery, monitoring and evaluation of the service in accordance with sustainability principles, including sustainability reporting into the organisation's strategic planning cycle (Table 10.5).

10.4.9 Creating Common Digital Platforms

Infrastructure services require the collaboration of many stakeholders and parties throughout the asset lifecycle. Efficient and effective communications are key challenges in infrastructure collaborations. However, the digital revolution has produced a range of solutions that can be effectively deployed to marshal key information through common digital platforms to enable timely and informed decision-making within the context of Sustainability Reporting Frameworks.

TABLE 10.5 The progressive ('value creation and capture') approach

The conventional ('value preservation') approach of considering sustainability in a water infrastructure service can be characterised as follows:	**The progressive ('value creation and capture') approach** of considering sustainability in a water infrastructure service can be characterised as follows:
The valuation of a water supply concession is predicated on the volume of water it can sell and at what margin. A typical water supply concession contract would focus on 'Goal 6: Ensure availability and sustainable management of water and sanitation for all', and a metric such as the following SDG indicator: '6.1.1 Proportion of population using safely managed drinking water services.	The valuation of a Water Resource Management Service is predicated on the utility value provided by the resource stewardship throughout the value chain and the socio-economic and environmental benefits arising from that resource stewardship. Mainstreaming of gender-inclusive demand assessments and stakeholder engagement have identified a diversified revenue base and quantified real market need. Moreover, this has achieved a fundamental respect in the community of the socio-economic value of the proposed service, materialised as a willingness-to-pay full for cost recovery for services achieving minimum performance standards. Capital costs could be lower through lower volume demand, through more efficient resource use. Costs of Capital are lower due to a demonstrably lower risk profile, with the following: The lower risk profile attracts lower insurance costs. The financiers that have taken the upfront risks in creating these stable revenue-generating business assets will have exit strategies of selling on the assets to locally based values-driven financing institutions such as bonds or pension funds, recapitalising the initial financiers to go on to repeat the process elsewhere. A Water Resource Management Service would recognise all the relevant SDGs that are touched upon, such as '**Goal 1:** End poverty in all its forms everywhere', '**Goal 3:** Ensure healthy lives and promote well-being for all at all ages', '**Goal 5:** Gender Equality', '**Goal 6:** Clean Water & Sanitation', '**Goal 8:** Decent Work & Economic Growth', '**Goal 9:** Industry, Innovation & Infrastructure', '**Goal 10:** Reduced Inequalities', '**Goal 11:** Sustainable Cities & Communities', '**Goal 12:** Responsible Consumption and Production', '**Goal 13:** Climate Action', '**Goal 14:** Life Below Water', '**Goal 15:** Life on Land', '**Goal 16:** Peace, Justice & Strong Institutions' and

(Continued)

TABLE 10.5 (*Continued*)

	The progressive ('value creation and capture') approach
	'**Goal 17:** Partnerships for the Goals'; indicators such as (the following is non-exhaustive, just exemplars) '**Target 1.4:** By 2030, ensure that all men and women, in particular the poor and the vulnerable, have equal rights to economic resources…', '**Target 3.9:** By 2030, substantially reduce the number of deaths and illnesses from hazardous chemicals and air, water and soil pollution and contamination', '**Target 5.5:** Ensure women's full and effective participation and equal opportunities for leadership at all levels of decision-making in political, economic and public life', etc.; and metrics such as '**Indicator 1.4.1:** Proportion of population living in households with access to basic services, '**Indicator 3.9.1:** Mortality rate attributed to unsafe water, unsafe sanitation and lack of hygiene (exposure to unsafe Water, Sanitation and Hygiene for All (WASH) services)', '**Indicator 5.5.2:** Proportion of women in managerial positions', etc.
The outcome is simply how do we stay in business in the short term.	The outcome is a business founded on sustainable principles that creates and captures value throughout the entire lifecycle of the asset. The business is more resilient to sustainability disruption. And through Sustainability Reporting is able to sustain the confidence of financiers to commit favourably priced finance throughout the asset lifecycle.

For infrastructure services, we need to incorporate specifications into all the project agreements that stipulate disclosure of information in the following two domains:

Enterprise Risk Management (ERM) is used by organisations for integrated management of their main business processes (including operations, accounting and resource management), in real time and mediated by software and technology (Institute of Risk Management, 2023). ERM systems can support automation and processes in finance, human resources, manufacturing, supply chain, services, procurement, *inter alia*. Thus, ERM systems can be readily augmented to deliver Sustainability Reporting compliant with Sustainability Reporting Standards.

Common Digital Data Environments[3] are digital information platforms that centralise project data storage and access and can feed both planned data (budgeted costs) and real measured data (actual costs) into an Enterprise Resource Planning. These can bring considerable efficiencies for engineering, procurement and construction projects through development of Digital Twins to optimise designs, integrating BIM workflows, then also facilitate efficient operations and maintenance of facilities through martialling real-time measured data captured and appraised against the Digital Twin.

An Application Programming Interface (API) is software programming based on established protocols that enables software applications to communicate with each other to exchange data, features and functionality (IBM, 2024). API solutions can be used to automatically (thus at minimal costs) pass on data in blockchains throughout value chains, whilst allowing all companies in that value chain to provide report on sustainability parameters both to their customers and also in their financial reporting, to their financiers.

10.5 SOME FINAL THOUGHTS

We cannot expect that simply using Sustainability Reporting in accordance with existing Sustainability Reporting Frameworks and Standards will bring about the behavioural changes necessary for our world to fall into a sustainable pathway. Indeed, there is already widespread abuse of these Sustainability Reporting tools to greenwash over excessive illth.

However, if we lobby for universally mandated obligations on governments and corporations to comprehensively measure and report within Sustainability Frameworks on the extent to which their activities are contributing or detracting from sustainability goals and also measure and report on how efficiently resources are being allocated in the provision of infrastructure services, **then we can at least understand with improved confidence on how far we are from a resilient and sustainable pathway**. Especially, it is the hope that **this enhanced visibility** will: (1) assert pressure on the wealthiest 1% that have significant control and influence on resource use to pivot their businesses towards sustainable business practices; (2) inform and empower our societies to more effectively petition governments to enact policies that shape our built environment around real sustainable needs as defined by the SDGs; (3) achieve sustainable outcomes through responsible consumption of infrastructure services (The approach described in this chapter should make a spectrum of information available, enabling consumers to make informed choices.); and (4) galvanise consumer pressure on corporates to plan for long-term financing of infrastructure over the entire asset lifecycle to continuously deliver sustainable services and outcomes.

Moreover, in the mature economies, there is generational transfer of wealth from baby-boomers to younger generations, whilst emerging economies are experiencing considerable wealth generation, as these economies mature. It is beholden on members of both groups to be active investors into financial products such as pensions and savings instruments that subscribe to Sustainable Reporting Frameworks and Standards and to hold the financiers to account.

We each then have the opportunity as consumers of our limited resources to take full responsibility for preventing illth and be smart in pivoting to the creation and stewardship of a 'resilient values-based Wealth of Nations' as a sustainable inheritance for passing onto our successive generations.

NOTES

1 Goodwill is an intangible value added as a premium into a business's valuation beyond the tangible assets less liabilities that is expected to bring additional future revenues such as client relationships, brand reputation or market expectations for intellectual property
2 Wider Economic Benefits are defined as the broader improvements in economic welfare arising from the provision of infrastructure, but which are not included in conventional economic assessments such as cost–benefit analysis (CBA). In economics terms these are characterised as arising from market imperfections where the prices of goods and services differing from costs to society as a whole.
3 A Common Data Environment (CDE) is a digital information platform that centralizes project data storage and access, typically related to a construction project and Building Information Modelling (BIM) workflows.

REFERENCES

ADB (2017) *Guidelines for the Economic Analysis of Projects*. Available at: https://www.adb.org/sites/default/files/institutional-document/32256/economic-analysis-projects.pdf

Australian Transport Assessment and Planning (2022) *Wider Economic Benefits*. Available at: https://www.atap.gov.au/tools-techniques/wider-economic-benefits

Collaborative Climate Action (2021) *Policy Brief Localising NDCs with Inspiration from the 2030 Agenda*. Available at: https://collaborative-climate-action.org/wp-content/uploads/Localising-NDCs-with-inspiration-from-the-2030-Agenda-Policy-Brief-GIZ.pdf

Equator-Principles (2020) *The Equator Principles – Environmental and Social Risk Management for Projects*. Available at: https://equator-principles.com/

Ftserussell (2025) *FTSE Russell*. Available at: https://www.ftserussell.com/about-us

Global Reporting Initiative (2013) GRI Index. Washington, DC: World Bank Group. Available at: https://documents.worldbank.org/curated/en/558111467993208712

GOV.UK (2021) *Duty of Care Waste Transfer Note Form*. Available at: https://www.gov.uk/government/publications/duty-of-care-waste-transfer-note-template

HM Treasury (2018) *Guide to Developing the Project Business Case. Better Business Cases for Better Outcomes*. Available at: https://assets.publishing.service.gov.uk/government/uploads/system/uploads/attachment_data/file/749086/Project_Business_Case_2018.pdf

IBM (2024) *What is an API (Application Programming Interface)*? Available at: https://www.ibm.com/topics/api

IFC (2012) *Policies and Standards*. Available at: https://www.ifc.org/en/what-we-do/sector-expertise/sustainability/policies-and-standards

IFRS (2024) *International Financial Reporting Standard Number 9*. Ifrs.org. Available at: https://www.ifrs.org/

Institute of Risk Management (2023) *What is Enterprise Risk Management*. Available at: https://www.theirm.org/what-we-do/what-is-enterprise-risk-management/

International Actuarial Association (2019) *Flood Risk Discussion Paper*. Resource and Environment Working Group. Available at: https://www.actuaries.org/IAA/Documents/Publications/Papers/REWG_Flood_Risk.pdf

MSCI (2022) *ESG Investing: ESG Ratings*. Available at: https://www.msci.com/our-solutions/esg-investing/esg-ratings

Rayworth, K. (2017) *Doughnut Economics*. Penguin Random House. Available at: https://doughnuteconomics.org/

Ruskin, J. (1910) *The Works of John Ruskin*, 30 vols. New York: T.Y. Crowell.

SASB (2019) *Sustainability Accounting Standards Board*. Available at: https://www.sasb.org/

Smith, A. (1776) *The Wealth of Nations*. London: W. Strahan and T. Cadell. Available at: https://www.earlymoderntexts.com/assets/pdfs/smith1776_1.pdf

Smith, A. (1812) *The Works of Adam Smith: The Theory of Moral Sentiments*. London: Penguin Classics, p. 494.

Sustainalytics (2024) *About U*s. Available at: https://www.sustainalytics.com/about-us

TCFD (2019) *Task Force on Climate-Related Financial Disclosures*. Available at: https://www.fsb-tcfd.org/

UN (2007) *CEO Water Mandate. Commitment to Water Stewardship*. Available at: https://ceowatermandate.org/

UNFCC (2023) *Measure Your Emissions*. Available at: https://unfccc.int/climate-action/climate-neutral-now/measure-your-emissions

UNFCCC (2015) *The Paris Agreement*. United Nations Climate Change. Available at: https://unfccc.int/process-and-meetings/the-paris-agreement

United Nations (2015) *The 17 sustainable development goals*. United Nations. Available at: https://sdgs.un.org/goals

World Bank (2016) *Environmental and Social Framework (ESF*). Available at: https://www.worldbank.org/en/projects-operations/environmental-and-social-framework

Summary and Conclusions

11

Scott Younger

11.1 KEY POINTS

It is not easy to predict the future sustainable development, certainly for the long term. A century ago, it would have been very difficult, if not impossible to predict where we would be today and even more so looking forward from the end of the 18th century, not least because the world then had only population of about a 1 billion, minimal industrialisation, almost zero CO_2 emissions and no hint of climate change as a looming hazard.

The benchmark of the Industrial Revolution has been chosen to represent when our ancestors ratcheted up the speed of development. A date quite commonly chosen to mark the start is 1783, when James Watt at Glasgow University identified an efficient way forward to better harness steam energy and initiate tremendous changes in our way of life. The downside, however, of the industrial age and its many benefits has been the great increase in pollution, damage to our green environments, diminishing resources and not least the potential future climate risks from carbon emissions.

The link between CO_2 emissions and industrialisation was demonstrated in Figure 4.3 which shows dramatically the change from near-zero emissions in 1750 (estimated at 9 million tons) and those today (38 billion tons); the very sharp rise in emissions in the last 100 years, particularly the past 60, being only interrupted for a short time by the Great Depression (1930–1938) and the COVID pandemic (2020–2021). The working definition used in this book requires that, to be classed as sustainable, development must include conservation of resources and protection of the environment for future generations. Current data, such as that shown in Figure 4.3, makes it difficult to argue that our current development is on a sustainable path. It should be noted that greenhouse gases (GHGs) are all inclusive carbon gases, e.g. methane, and not just CO_2, as discussed in Chapter 4.

DOI: 10.1201/9781003566274-11

Sustainable development only appeared as a term about 33 years ago. At the same time, it was realised that the computer had advanced sufficiently, from the 1950/1960s, that more complex problems could be tackled, and people started to see if they could attempt to solve huge problems with large datasets that had previously been considered impossible. It started a trend for trying to forecast sustainable trends. For a while, sustainable development only considered three issues, social, economic and environmental. Experience showed that the political situation had most often to be considered as well, which gave rise to the SEEP equation – Social, Economic, Environmental, Political (Younger et al., 2012). This is the underlying theme of the early chapters in the book.

After an opening chapter entitled "Setting the scene", this book provides an overview of governance and problems of corruption in Chapter 2, followed by discussion on the importance of education and the often-linked never-ending issue of poverty in Chapter 3. Chapter 4 which deals with the properties of the landscape and leads up to what has become an issue of today, climate change. Chapter 5 follows, which discusses the environment in the context of sustainable development.

Despite significant progress in regulatory reform and related compliance measures over the past 25 years, corruption risks remain high in relation to the planning and provision of public infrastructure. Not only does this aggravate the infrastructure funding gap, but related shortcomings in governance and quality management also undermine efforts to strengthen associated sustainability considerations. Though there is no single, or simple solution, there is growing evidence that corruption risks can be better understood and mitigated through approaches that are instead characterised different aspects of recognised good practice.

It seems inevitable that poverty would continue to be around affecting humanity in many places in the world for the foreseeable future, negatively impacting on sustainable development. Lessons from case studies seem to indicate that in virtually all cases where poverty was found, there was a solution, not just money, but most of all targeted help and attitudes to tackle the problem, with self-help and instilled confidence being the ultimate aim.

Unlike corruption, poverty can be eradicated entirely in defined areas or situations, and we must strive to widen our ambitions. The resistance to elimination of poverty in some situations is authoritarian, selfish, often brutal rule which cares for its self-preservation above all else. There are still too many countries which exhibit that tendency, which, alas, hold back poverty alleviation and its part in overall sustainable development.

Earth is constantly changing in geological time, which generally allows us to make the necessary adjustments For example, a barrage had to be built across the Thames estuary a few decades ago to protect the city of London from being inundated with water from combinations of high tides and storm surges. It is a reminder that the growing size of degraded land is not just arid desert, but is found also as green areas, intensively farmed or stripped of natural vegetation, or rainforests devasted by uncontrolled logging. With human population still expanding, the extent of the problem is only going to get bigger and really requires our attention now.

The Intergovernmental Panel on Climate Change (IPCC), the agency for assessing the science related to climate change, was created by the UN environment programme and the World Meteorological Organisation in 1998, to provide policymakers with regular assessments on climate change, its implications and potential future risks, as well as to put forward adaption and mitigation options. The predictions of the IPCC are based on average global surface temperature, a necessary condition because of the vast amount of computing power required for climate modelling.

However, humanity, along with plants and animals, does not live in a world of average air temperature, but in various regions of the world that experience a great variety of temperature, rainfall and wind patterns. These are driven by a combination of celestial mechanics, solar radiation, atmospheric physics, the rotation of the Earth, ocean circulation, the latitude of the region and concentration of GHGs. Hence, the IPCC's long-term forecasts are fundamentally of little point, and a combination of adaption and mitigation would be more apposite. There is some high-level political controversy over the IPCC work, although it must be applauded for drawing attention to the possible negative impacts of climate change.

From the work of Koppen and Geiger, the world is presented as 32 climatic regions which are defined in terms of their topography and flora as well as climate. Taking this along with the population, both in terms of total and areal growth, how much rural landscape against how much urban, it is deemed possible to better plan the improvements needed in the various regions. This top-down approach could be supported at more local level and would take into account the levels of education, poverty and development.

It is considered that there are three issues that are essential for human life, water, energy and to a lesser extent, transport. These topics are covered in Chapters 6–8. In Chapter 6, after a summary of fossil fuels, the focus is on the forms of nuclear energy. Nuclear is given full coverage, particularly since the cleaner new fusion is expected to replace well-tried but radioactive fission within a decade or so. Renewable sources of energy allow the world several options to steer away from fossil fuels and thus reduce GHG emissions. Energy is now capable of being sourced from a range of renewable options, but fossil fuels will continue to be used by some nations, which have little option or are sceptical of the argument for climate change.

While hydro energy could be eventually overtaken by wind and solar in some cases, it will continue to play a key part to back up variable renewables, although increasingly erratic rainfall due to climate change is also disrupting hydro production in many parts of the world. A guideline target for maximum use of fossil fuels would seem to be below a level that was used pre-1960s, which indicates that emission levels should be not much more than pre-industrial ones, when no obvious changes were taking place to the climate. Table 11.1 together with its back-up data (Ritchie, 2025) illustrates key points in global emission trends.

1. The significant drop off in the UK percentage, together with Europe but a less dramatic decrease by USA.
2. Major increase in percentage contribution by China.
3. Seven emitters account for only 63% of emissions – as opposed to nearly 100% up to end of 19th century.
4. 47% of emissions come from a wide spread of countries globally, with 20% from Asian countries outside China and India.
5. Some countries have taken serious measures towards halting increasing carbon emissions if not reversing the unwanted trend, and the signs are there that they have been successful.
6. Smaller, relatively undeveloped countries hardly commit any carbon emissions individually but as a group contribute 17% to the total.

TABLE 11.1 Worldwide percentage of CO_2 emissions for selected countries (Ritchie, 2025)

	YEAR (% CO_2)						
COUNTRY	1750	1800	1850	1900	1950	2000	2023
UK	100	96	57	21	10	1	1
Europe	0	3	29	36	23	14	1
USA	0	1	13	34	42	24	13
China	0	0	0	0	1	14	32
India	0	0	0	1	1	7	8
Russia	0	0	1	2	7	6	5
Japan	0	0	0	1	2	5	3
Total %	100	100	100	95	86	71	63

Water is vital and without continuing resources, humanity would decline and die out. It is found in most places: as seawater in the oceans, 97% of the total, and as freshwater appearing as groundwater, rivers, lakes and rainfall, just 3%. The majority of the freshwater is found as groundwater and, with increasing population, there is a growing deficit now appearing in groundwater, which is of concern. Freshwater is not evenly distributed, and it is subject to the regional climate. Pollution and poor sanitation are not adequately considered and have increased markedly with the population. Virtually all the uses of water require the construction efforts of civil engineers to bring it to urban localities, in particular, and safely deal with the waste afterwards. Water and its uses will be a priority issue as the century unfolds as well as in the future thereafter (Younger, 2025).

As discussed in Chapter 7, different parts of the water sector are struggling against the rise in population. This is a sector that requires a good understanding of all aspects of water engineering, although it is quite common to find specialisation of specific branches – irrigation, water supply, sanitation, and in fields such as hydropower. A fairly recent finding is that, as the population has been expanding, there has been greater use of underground reserves, such that there has been a drop in the level of the water that is stored underground. This has to be addressed and is seen as one of the major problems to be faced in the coming decades, and civil engineers will be called upon to find reliable solutions.

Key issues of sustainable transport are addressed in Chapter 8, and in Chapter 9, comments are given on other civil engineering issues. Transport has been at the centre of our development from the earliest days, but an increasing price has been paid for developments in the last two centuries in terms of damage to our environments on land, sea and atmosphere. The design of roads, whether urban or rural, depends on the traffic evaluation.

Transport is a target for criticism with respect to GHG emissions. Globally, 95% of the world's transport energy is still coming from fossil fuels. In the future, transport will have to be less focussed on speed, size and national pride and more on efficiency and the use of clean or renewable energy. There has been a very significant growth in traffic in all its forms in the past 60 years, albeit vehicles are still largely propelled by fossil fuels.

The advent of the computer allowed the subject of traffic engineering to become a main subject for study at university.

Chapter 10 has made recommendations on the issues around understanding and assessing the financial sustainability of civil engineering projects. Current financial decision-making is predominantly based around a retrospective review of financial performance reporting that deprives decision makers of insights from longer-term datasets. Proposals are outlined for mainstreaming sustainability into future business practices. A selection of sustainability financial reporting tools is recommended in the context of working towards SDGs.

11.2 CONCLUSIONS

There are other matters, apart from those which stimulate data engineers and scientists, whose world is really just beginning, with the exploration of space offering endless amazing opportunities. There is the management or governance of what is a growing human population. One of the first priorities for any government is going to be to provide shelter and food for the current and forecast populations. For the next half century at least, many governments will be struggling to provide shelter, particularly for the poorer sections of society. Furthermore, there will be a critical need for accommodation in the fast-growing countries of Sub-Saharan Africa, particularly in the next few decades. Measuring the provision of housing in terms of sustainability by 2100, it is likely that there will still much to do, partly because of further conflicts, which see no end of stopping.

There is one area of the human experience that needs continual airing, i.e. science and technology, including medical sciences. If the past experience, since the start of the Industrial Revolution, is meaningful, then this aspect provides most confidence and hope for the Future. The recent awakening to the vast range of possibilities of artificial intelligence (AI) means that the question of how sustainable development is measured will differ in the decades to come. It is a very useful advancement if only to assess very large problems, larger than we can evaluate the impact of those today. It can also help satisfy our inherent curiosity and sense of adventure in terms of not only planet Earth, but the Moon, the Solar System and star systems further afield.

Two issues which are increasing in prominence today are the expansion in the development of electric vehicles (EVs) and, secondly, AI. With regard to EVs, as has been said, the Scandinavian countries have shown the most interest but China is committed to become the most important player in this market, despite the large involvement in fossil fuel (coal) energy. Guangzhou in south China is a large city but the streets are relatively quiet and unpolluted with the significant numbers of EVs.

The AI developments are astonishing, and with the development of very large computers, we are going to witness some solutions that were only a dream six decades ago, as we then marvelled at the first main frames to come to universities and research institutions, being superseded in a matter of two to three decades when many problems could be handled more efficiently by a desktop, miniaturisation, and in a few minutes rather than taking 2–3 days within two decades of further development. The Digital Age had arrived, and soon will be taken over by AI, changing the way that sustainable development for the future years will be assessed, probably quite different from the past but at the same time not forgetting our duty to those struggling with poverty.

One final thought. Friedman (2009), the Founder and CEO of Stratfor, the US's leading international affairs intelligence company, made a forecast for this 21st century. He acknowledged at the start that it would be difficult and some details would be wrong. However, we can take some of the forecasts as correct, namely the advances in space and the impact that this is having on how warfare will change in the more developed countries. For instance, the current fast development of drones and the ideas that, in a matter of a few decades, platforms may be created in space operated by robots but controlled by humans on Earth, providing an all-seeing eye. The US would be the most advanced in this, developing through NASA's huge fund of space knowledge, but other countries with a space programme will attempt to vie with this. Some ructions will ensue, possibly leading to a major war in a few decades when all these advances will be tested.

Sustainable development will be significantly challenged and will have to be reviewed in a different light in years to come. Some nations, while continuing with the necessary attention to water and energy and transport, will have to learn how to govern with a portfolio which could have a space element. In time, towards the end of the century, the US will possibly have declined somewhat in power as other leading countries come to the fore and share the top position, as history as shown with the empires of old Empires may last a century and more but never in perpetuity. Things may change in the future with the fast turn round of ideas, some building on those that came before. It is likely that the forms of governance will also change as more countries mature and cooperate. Some of the ways in which sustainable development will be examined will have to change to reflect better the different times; a matter of a few decades.

BREAKING NEWS. Rock Springs, USA, becomes the first landfill in the World to divert electronic waste using a pyrolysis process. The next plant will be in UK in 3rd quarter, 2026.

REFERENCES

Friedman, G. (2009) *The next 100 years*. Allison & Building sby, London.

Ritchie, H. (2025) *Who Emits the Most CO_2 Today?* OurWorldinData.org. Available at: https://ourworldindata.org/annual-co2-emissions

Younger, J. S., Booth, D. J., and Kurniawan, K. (2012) Sustainable development: East Bali Poverty Project. *Proceedings of the Institution of Civil Engineers – Municipal Engineer* 165(1):43–58.

Younger, J. (2025) Briefing: The future of water: a looming crisis, attention required. *Proceedings of the Institution of Civil Engineers – Civil Engineering*, 178 (4). https://doi.org/10.1680/jcien.25.00175.

Index

Note: **Bold** page numbers refer to tables; *italic* page numbers refer to figures and page numbers followed by "n" denote endnotes.

For Product Safety Concerns and Information please contact our EU representative GPSR@taylorandfrancis.com Taylor & Francis Verlag GmbH, Kaufingerstraße 24, 80331 München, Germany